Dahlem Workshop Reports
Life Sciences Research Report 21
Animal Mind – Human Mind

The goal of this Dahlem Workshop is:
to explore the nature of the animal mind
and to develop new approaches to its understanding

Life Sciences Research Reports
Editor: Silke Bernhard

Held and published on behalf of the
Stifterverband für die Deutsche Wissenschaft

Sponsored by:
Senat der Stadt Berlin
Stifterverband für die Deutsche Wissenschaft

Animal Mind – Human Mind

D. R. Griffin, Editor

Report of the Dahlem Workshop on
Animal Mind – Human Mind
Berlin 1981, March 22-27

Rapporteurs:
M. Dawkins · W. Kintsch · H. J. Neville · R. M. Seyfarth

Program Advisory Committee:
D. R. Griffin, Chairman · J. F. Bennett · D. Dörner
S. A. Hillyard · B. K. Hölldobler · H. S. Markl · P. R. Marler
D. Premack

Springer-Verlag Berlin Heidelberg New York 1982

Copy Editors: M. Cervantes-Waldmann, K. McWhirter
Photographs: E. P. Thonke

With 4 photographs, 30 figures, and 2 tables

ISBN-13:978-3-642-68471-5 e-ISBN-13:978-3-642-68469-2
DOI: 10.1007/978-3-642-68469-2

CIP-Kurztitelaufnahme der Deutsachen Bibliothek:

Animal mind – human mind: report of the Dahlem Workshop on Animal Mind – Human Mind, Berlin 1981, March 22-27 / D. R. Griffin, ed. Rapporteurs: M. Dawkins ... [Dahlem Konferenzen. Held and publ. on behalf of the Stifterverb. für d. Dt. Wiss. Sponsored by: Senat d. Stadt Berlin; Stifterverb. für d. Dt. Wiss.]. – Berlin; Heidelberg; New York: Springer, 1982.

(Life sciences research report; 21)

(Dahlem workshop reports)

NE: Griffin, Donald R. [Hrsg.]; Dawkins, Marian [Mitverf.]; Workshop on Animal Mind, Human Mind <1981, Berlin, West>; Dahlem Konferenzen; GT

Softcover reprint of the hardcover 1st edition 1982

2131/3014 – 5 4 3 2 1 0

Table of Contents

The Dahlem Konferenzen

DIRECTOR:
Silke Bernhard, M.D.

FOUNDATION:
Dahlem Konferenzen was founded in 1974 and is supported by the Stifterverband für die Deutsche Wissenschaft, the Science Foundation of German Industry, in cooperation with the Deutsche Forschungsgemeinschaft, the German Organization for Promoting Fundamental Research, and the Senate of the City of Berlin.

OBJECTIVES:
The task of Dahlem Konferenzen is:

to promote the interdisciplinary exchange of scientific information and ideas,

to stimulate international cooperation in research, and

to develop and test different models conducive to more effective scientific meetings.

AIM:
Each Dahlem Workshop is designed to provide a survey of the present state of the art of the topic at hand as seen by the various disciplines concerned, to review new concepts and techniques, and to recommend directions for future research.

PROCEDURE:
Dahlem Konferenzen approaches internationally recognized scientists to suggest topics fulfilling these criteria and to propose members for a Program Advisory Committee, which is responsible for the workshop's scientific program. Once a year, the topic suggestions are submitted to a scientific board for approval.

TOPICS:
The workshop topics should be:

of contemporary international interest,

timely,

interdisciplinary in nature, and

problem-oriented.

PARTICIPANTS:
The number of participants is limited to 48 for each workshop, and they are selected exclusively by a Program Advisory Committee. Selection is based on international scientific reputation alone and is independent of national considerations, although a balance between Europeans and Americans is desirable. Exception is made for younger German scientists for whom 10% of the places are reserved.

THE DAHLEM WORKSHOP MODEL:
A special workshop model has been developed by Dahlem Konferenzen, the *Dahlem Workshop Model*. The main work of the workshop is done in four small, interdisciplinary discussion groups, each with 12 members. Lectures are not given.

Some participants are asked to write background papers providing a review of the field rather than a report on individual work. These are circulated to all participants 4 weeks before the meeting with the request that the papers be read and questions on them formulated *before* the workshop, thus providing the basis for discussions.

During the workshop, each group prepares a Report reflecting the essential points of its discussions, including suggestions for future research needs. These reports are distributed to all participants at the end of the workshop and are discussed in plenum.

PUBLICATION:
The Dahlem Workshop Reports contain:
- the Chairperson's introduction,
- the background papers, and
- the Group Reports.

Animal Mind - Human Mind, ed. D.R. Griffin, pp. 1-12.
Dahlem Konferenzen 1982. Berlin, Heidelberg, New York: Springer-Verlag.

Introduction

D. R. Griffin
The Rockefeller University
New York, NY 10021, USA

Prominently displayed in each meeting room was the stated purpose of this Dahlem Conference: To explore the nature of the animal mind and develop new approaches to its understanding. We were inquiring whether nonhuman animals have mental experiences, and if so which animals, under what conditions.

We were also trying to develop ways to learn what any mental experiences that may occur are actually like to the animals concerned. Do they ever think about images or representations of the outside world, examine them, and make choices as to how their own behavior might change the situation? Or are all nonhuman animals invariably in a state comparable to a human sleepwalker, responding mechanically, even though often adaptively?

In recent years such questions have been taboo in the behavioral sciences. But, as I have suggested elsewhere (7, 8), it may now be time to lay these inhibitions aside and make a new attempt to explore important scientific territory that has been largely neglected for several decades. Many philosophers such as Armstrong (1), Bunge (2), Dennett (3), Fodor (6), and Taylor (11) have long since escaped from the counterproductive self-paralysis that was once advocated by positivists and strict behaviorists. Although reopening these questions may seem pointless or even harmful to died-in-the-wool behaviorists, there is no longer any reason, if there ever was, to hold back from inquiries that might be fruitful by denying, in advance, hypotheses about both human and animal minds that may in fact be correct and significant. Total certainty is not necessary as we take preliminary steps to reduce our ignorance in this area. The stakes are high, nothing less than the scientific understanding of what it is like to be animals of various kinds and what is truly unique about the human mind.

I feel it is of great importance that we recognize at the outset how little we really know about animal minds and make the psychologically difficult and frustrating effort to act as humble, inquiring investigators who are seeking to learn, rather than merely reiterating dogmatic assertions that are comfortably familiar but often based on fragile and ambiguous foundations. The participants in this conference have looked ahead as open-minded agnostics and concentrated on devising new approaches to these long neglected questions. In the published papers and state of the art reports we now have available a variety of thoughtfully considered reviews together with a number of enterprising suggestions for future investigations.

At times we all tend to rule out a priori the possibility that various kinds of mental experiences could occur in one or another group of animals. For instance, reflexive self-awareness, thinking about one's own thoughts, is widely asserted to be a unique human capability, and any sort of intentional communication is vigorously denied to insects on the ground that they are genetically programmed robots. But then we asked ourselves whether this degree of confident certainty was really appropriate for inquiring scientists dealing with areas where the available data are clearly inadequate. We tend to think that were we in the animal's skin we would immediately see the folly of retrieving beer cans into a bird's nest or reiterating preliminary sequences in an insect's behavioral repertoire when a simple shortcut would be more efficient. Or we imagine that we would ignore the oleic acid on a live and wriggling sister or mother and refrain from evicting her from our hive. But does the occurrence of unintelligent behavior suffice to demonstrate the total absence of mental experience under any circumstances? Ethologists from some distant galaxy could easily discern examples of stupid and maladaptive behavior in our own species. But do instances of human stupidity prove that none of us is ever consciously aware of what he is doing? No available evidence compels us to believe that insects, or any other animals, experience any sort of consciousness, or intentionally plan any of their behavior. But neither are we compelled to believe the contrary. In areas where data are few and of limited relevance, dogmatic negativity can easily limit what scientists even try to investigate, and thus perhaps delay or prevent important insights and discoveries.

Many of the participants agreed that a good starting point would be to consider what we know of our own thinking, subjective feelings, and consciousness, and then move on to inquire whether other species experience anything similar. Such an approach was once considered fallaciously anthropomorphic. But it seems now to be widely if not universally recognized that this is a serious objection only if one has already assumed in advance that conscious thinking <u>is</u> uniquely human, and the accusation of anthropomorphism is then merely a reiteration of the prior conviction. Animal experiences may differ widely from ours, but the initial task is to detect their existence. We can then begin to learn, if we are fortunate, how similar or different they may be. It has been pointed out by several participants that conscious experience is not an all-or-nothing phenomenon. In our own species it occurs in a variety of forms and degrees, so that it is plausible to suppose that there may be at least as great a range of variability in its nature, content, and complexity in nonhuman animals. If and when mental experiences do occur in various animals, they will probably not be elegantly balanced, scholarly judgments; they may well be highly emotional, imperfect, even mistaken or superstitious.

One basic point of considerable importance is to distinguish clearly between awareness and responsiveness. It is obvious in our own case that either can occur without the other, and the same may well be true for other species. Awareness entails the experience of thinking about something, and ordinarily this experience includes some evaluation or recognition of relationships between the objects and events about which we are thinking. Granting all the widely recognized limitations of introspection, we do know something about our own thoughts and feelings, as emphasized by Taylor (11). At the very least we often know, and can report, what it is we are consciously thinking about. When and where, if at all, do comparable experiences occur in other species?

Many behavioral scientists use the terms mind, mental, and cognition to include information processing or problem solving that need not be accompanied by consciousness or subjective feelings. Some of the papers in this volume deal almost entirely with studies of human thinking. Others deal with information processing, learning, problem solving or social manipulation in animals, but without any expressed concern for possible mental experiences that might accompany the behavior under consideration. These papers provide background material for our ultimate objective: to understand and compare mental experiences of animals and men.

While scientists quite properly attempt to define whatever terms or concepts they use, precise definitions of mental processes and experiences are almost impossible for the simple reason that we know so little about the phenomena we seek to define. With better understanding, definitions may improve. To quote Woodfield (13): "If the term being defined is vague or open-ended, then a definition will be exact only if it too is vague and open-ended." Although philosophers have been deeply concerned with the nature of human minds, they seldom commit themselves to the sort of specific definitions that scientists find satisfying. Many definitions are circular, in that one term is used to define another and vice versa. For example, the Oxford English Dictionary (Compact edition, Oxford University Press, 1971) devotes several pages to the definitions and usage of terms dealing with mental experience. From these I have selected a few major ones that seem especially appropriate to the questions about the possible existence of animal minds:

Mind: Thought, purpose, intention; the action or state of thinking about something; the seat of a person's consciousness, thoughts, volitions, and feelings; the system of cognitive and emotional phenomena and powers that constitute the subjective being of a person.

Aware: Watchful, vigilant, cautious, on one's guard; informed, cognizant, conscious, sensible.

Intend: To bend the mind to something to be done, to purpose, design, mean to.

Conscious: Aware of what one is doing or intending to do, having a purpose and intention in one's actions.

Consciousness: The totality of the impressions, thoughts, and feelings, which make up a person's conscious being.

Feeling: Consciousness of (a subjective fact); pleasurable or painful consciousness, emotional appreciation or sense (of one's own condition or some external fact).

Think: To conceive with the mind; to have in the mind as a notion, idea, etc.; to be of opinion, deem, judge, etc.

One textbook of philosophy (5) defines mind as follows: "Feelings, sensations, dreams, and thoughts are the sort of phenomena which are usually classified as mental." Similar definitions, implicit or explicit, are articulated in the articles on consciousness, intention, thinking and the mind-body problem in Edwards' Encyclopedia of Philosophy (4). On the other hand some definitions are quite specific. Kenny advanced as a criterion of mind "to have the capacity to acquire the ability to operate with symbols in such a way that it is one's own activity that makes them symbols and confers meaning on them" ((10), p. 47). Longuet-Higgins stressed the capability for intentional action: "An organism which can have intentions I think is one which could be said to possess a mind (provided it has)...the ability to form a plan, and make a decision - to adopt the plan" ((9), p. 136).

Armstrong (1) argues that minds can be encompassed into a strictly materialist or physicalist philosophy, and that they are best viewed as states or processes in the central nervous system that are potentially capable of producing behavior, that is, thoughts are viewed as dispositions to behave in certain ways. But we are aware of only a small fraction of the processes occurring in our nervous systems that produce, or at least strongly affect, our behavior. In fact, this definition of a mind would seem to include all physiological processes that serve to initiate any sort of observable behavior in any living organism. To Armstrong's definition I would add the capability of thinking about objects and events, some of which may be remote in time or space, for example, an animal's remembering yesterday's tasty food or anticipating the warm dry nesting chamber at the end of its burrow, and our imagining the Battle of Gettysburg or the first manned landing on Mars.

Armstrong goes on to define consciousness as perception of one's own mental states. He illustrates this by citing as an example automatic driving by a person who drives an automobile effectively but without conscious thought about the process: "an account of mental processes as states of the person apt for the production of certain sorts of behaviour very possibly may be adequate to deal with such cases as that of automatic driving. It may be adequate to deal with most of the mental processes of animals, which perhaps spend most of their lives in this state of automatism." Armstrong defines true consciousness as "perception or awareness of the state of our own mind... a self-scanning mechanism in the central nervous system."

Another materialist philosopher, Bunge (2), begins a book about the nature of minds by stating that "Perceiving, feeling, remembering, imagining, willing, and thinking are usually said to be mental states or processes. (We shall ignore...the quaint view that there are no such facts.)" Bunge goes farther than most philosophers in recognizing that at least some mammals and birds experience mental states and processes, but he asserts (pp. 74-75) that "All and only the animals endowed with plastic (uncommitted or not wired-in) neural systems are capable of being in mental states (or undergoing mental processes)." Bunge rigidly denies any mental life to all poikilothermic vertebrates and to all invertebrates, ignoring abundant evidence that in many cases their behavior is plastic and versatile. But the following quotations and paraphrases of his definitions and logical criteria are helpful guides to our thinking about animal minds:

"Mental events can cause nonmental events in the same body and conversely...(p. 163). We take it for granted that animals of several species know how to perform certain actions, know some constructs, and have some knowledge of events. We include among the latter the empathic knowledge of other animals. Empathy, extolled by intuitionists and mistrusted by rationalists, is admittedly fallible - but it is also indispensable.

An animal is aware of (or notices) stimulus x (internal or external) if and only if it feels or perceives x. The animal is conscious of brain process x in itself if and only if thinks of x....The consciousness of an animal is the set of all the states of its CNS in which it is conscious of some neural process or other in itself.

An animal act is voluntary (or intentional) if and only if it is a conscious purposeful act....An animal acts of its own free will if and only if its action is voluntary, and it has free choice of its goal(s)....All animals capable of being in conscious states are able to perform free voluntary acts.... An animal (i) has (or is in a state of) self-awareness if and only if it is aware of itself (i.e., of events occurring in itself) as different from all other entities; (ii) has (or is in a state of) self-consciousness if and only if it is conscious of some of its own past conscious states; and (iii) has a self at a given time if and only if it is self-aware or self-conscious at that time."

Thus Armstrong, Bunge, and many others distinguish between awareness and consciousness on the basis that awareness is a sort of perception of an object or event, which may be known through current sensations, remembered from the past or anticipated for the future, but consciousness is considered to require in addition a special kind of self-awareness. This special self-awareness is not simply awareness of some part of one's own body or some process occurring within one's skin, but a propositional awareness that it is _I_ who am thinking or feeling something or other, that _I_ am the organism aware of this or that object or event. This usage is not universally followed in describing human mental experiences, and it is of course even more difficult in the case of other animals to gather evidence bearing on this distinction. But the distinction is an important one which cognitive ethologists should keep in mind and try to develop methods capable of discriminating perceptual awareness of objects and events from self-conscious awareness on an animal's part that it is itself experiencing the awareness under consideration. Signing apes might provide some pertinent data insofar as they can be shown, convincingly, to sign about their own memories, feelings, or desires and distinguish these from the memories, desires, or feelings of others. And we should ask ourselves whether it is really plausible to suppose that an organism can experience any reasonably full spectrum of perceptual awareness without some notion that it, itself, is the perceiver.

In struggling with the thorny questions of definition, I keep coming back to the self-evident fact that consciousness involves thinking about objects and events which may or may not be eliciting immediate responses and may or may not be closely linked to current sensory input. Scanning many sorts of internal representations and comparing them with each other and with current perceptions or anticipations of future events seem to me close to the heart of our intuitive ideas about our own consciousness. How, if at all, can we judge whether other species think about objects and events?

In our own minds the past and future may be more or less remote from the present, so that one important aspect of human consciousness is the existence of a time scale of some sort from the very remote past though the immediate present and extending varying degrees into the future. Although we may assume, in the absence of any evidence to the contrary, that we are the only species capable of contemplating events occurring before our own birth or after our death, we must recognize that any animal with reasonable capabilities for learning and memory has a limited range of past experiences on which it can call. It is more difficult to detect an animal's projections into the future, but at least short-range intentions to perform particular actions, or expectations that a given event or sensation is likely to occur, seem quite plausible. As emphasized by Tolman (12) the apparent surprise and puzzlement shown by animals when something they have every reason to expect fails to appear provides strong evidence that they did indeed expect it.

At the outset of the conference I suggested two particular approaches that seemed especially promising and which I hoped would receive thoughtful attention. The first is my optimistically favorite notion that animal communication can provide us with a useful window on animal thinking or feeling in a fashion analogous to the way in which human language and nonverbal communication tell us most of what we know or believe about the thoughts and feelings of other people. As I have suggested elsewhere (7, 8), it seems possible that ethologists interested in animal cognition could learn a great deal through experimental dialogues in which an experimenter communicates back and forth with an animal to inquire about its mental experiences. This approach received only passing attention in our discussions, and it remains unexplored scientific territory waiting for adventurous pioneers.

A second approach which is readily compatible with our traditions of empirical research would be to inquire whether event-related potentials such as the "P 300" or other physiological measures of brain function might provide objective indices of conscious thinking. Whatever sort of brain activity produces human consciousness must entail some pattern of neuronal activity. Most neuronal activity involves electrical potentials, so that in theory there must be electrical (or perhaps chemical) correlates of conscious thought. Event-related potentials are discussed in the two chapters by Hillyard and Bloom and by Neville and Hillyard. Some of them occur when a stimulus which a human subject is anticipating suddenly occurs in a monotonous series of other stimuli. Others seem related to expectancy, recognition, and decision making. Are any of these event-related potentials uniquely correlated with consciousness in human subjects? If so, perhaps the occurrence of potentials with similar properties in other species could provide objective evidence of conscious thinking on their part. If not, future refinements of neurophysiological investigation may disclose other objective signs of conscious thinking. This general approach is motivating neurophysiologists to explore these exciting possibilities. Even if they cannot develop a reliable "litmus test" of conscious awareness, thorough and quantitative analyses of the relationships between various types of event-related potentials in men and animals under conditions when mental experiences seem likely will surely throw considerable light on the fundamental questions addressed at this conference.

I firmly believe that it has been worthwhile to explore these highly significant though extremely thorny problems. If the progress seems limited, we should recall the difficulty of what this conference was attempting to do. We set our sights so high that we should be satisfied with small but open-ended beginnings. I believe that most of the participants became more interested in planning investigations designed to throw light on animal minds. Scientific progress results from a complex nexus of interacting influences, and no one conference can realistically hope to achieve unaided as important a new departure as I hope we will see in future years. But the discussions in Berlin and the publication of this volume will surely contribute substantially to the development of an effective area of investigation which I like to call cognitive ethology. Of course other terms may be equally applicable; the important question is not what labels we use, but the degree to which we can come to understand animal and human minds.

Finally I should like to take this opportunity to express my appreciation, indeed my admiration, for the thoughtful work which all participants in this conference and the hard-working staff of the Dahlem Konferenzen have devoted to attempting what many of our colleagues have believed to be a hopeless task. To Silke Bernhard we participants all owe a great debt for her imaginative enterprise in planning this particular conference and making our stay in Berlin so pleasantly stimulating. I hope readers of this book will be able to share our excitement in wrestling with embryonic ideas about new approaches to these fascinating questions.

REFERENCES

(1) Armstrong, D.M. 1980. The Nature of Mind, and Other Essays. St. Lucia, Qld., Australia: University of Queensland Press, and Ithaca, NY: Cornell University Press.

(2) Bunge, M.A. 1980. The Mind-Body Problem, A Psychobiological Approach. New York: Pergamon.

(3) Dennett, D.C. 1980. Brainstorms, Philosophical Essays on Mind and Psychology. Montgomery, VT: Bradford Books.

(4) Edwards, P., ed. 1967. Encyclopedia of Philosophy. New York: MacMillan.

(5) Edwards, P., and Pap, A., eds. 1973. A Modern Introduction to Philosophy, 3rd ed. New York: MacMillan.

(6) Fodor, J.A. 1975. The Language of Thought. New York: Thomas Y. Crowell.

(7) Griffin, D.R. 1978. Prospects for a cognitive ethology. Behav. Brain Sci. 1: 527-538, with multiple commentaries pp. 555-629.

(8) Griffin, D.R. 1981. The Question of Animal Awareness, 2nd ed. New York: The Rockefeller University Press.

(9) Kenny, A.J.P.; Longuet-Higgins, H.C.; Lucas, J.R.; and Waddington, C.H. 1972. The Nature of Mind. Edinburgh: Edinburgh University Press.

(10) Kenny, A.J.P.; Longuet-Higgins, H.C.; Lucas, J.R.; and Waddington, C.H. 1973. The Development of Mind. Edinburgh: Edinburgh University Press.

(11) Taylor, J.C. 1962. The Behavioral Basis of Perception. New Haven, CT: Yale University Press.

(12) Tolman, E.C. 1932. Purposive Behavior in Animals and Men. New York: Century.

(13) Woodfield, A. 1976. Teleology. London: Cambridge University Press.

Animal Mind - Human Mind, ed. D.R. Griffin, pp. 13-32.
Dahlem Konferenzen 1982. Berlin, Heidelberg, New York: Springer-Verlag.

Brain Functions and Mental Processes

S. A. Hillyard* and F. E. Bloom**
*Dept. of Neurosciences, M-008, University of California, San Diego, La Jolla, CA 92093
**The Salk Institute, La Jolla, CA 92037, USA

Abstract. Inferences about animal minds have typically been based upon the behavioral repertoire and ecological adaptations of the species in question. Supporting evidence can be obtained from cross-species comparisons of brain structure, function and organization, particularly comparisons between the brains of animals and humans. Here we describe several brain systems and functions which appear to be linked with mental processes in man and might, therefore, serve as markers for homologous or similar processes in animals.

INTRODUCTION

The study of mental processes - perceptions, thoughts, intentions, and the like - is difficult to accomplish in any species, man included. Since we only have access to a person's verbal reports and other behavioral outputs, we generally get an indirect and fragmentary picture of their mental activities, even in carefully controlled experimental situations. This problem is greatly compounded when we attempt to understand the mental processes of animals. In the absence of verbal fluency, we have to approach an animal's mind by examining the many facets of its behavioral repertoire including intraspecific communications, social interactions, learning and problem-solving abilities, and ecological adaptations.

In this paper we consider another avenue towards the comparative analysis of mental processes - the examination of cross-species similarities in brain organization and function. The goal of this approach is first, to determine which brain activities are essential for different kinds of mental activities and then to chart their incidence across the animal kingdom. The emphasis here will be on identifying those brain systems and functions which have been linked (however tentatively) with particular mental activities in man. The existence of similar brain activities in nonhuman species might then be taken as prima facie evidence for a similarity of mentation.

An underlying assumption being made here is that mental phenomena are strictly dependent upon, and in a sense, are equivalent to specific spatio-temporal configurations of brain activity. Not only are different types of mental processes translatable into distinct patterns of brain activity, but the presence of one may be inferred from the other. Such monistic assumptions about minds and brains are commonly made by physiological psychologists. The problem, of course, is that we are not sure which levels or aspects of brain structure and function are critical for most kinds of mental events. Our current predilection is to propose cross-species comparisons of integrative brain functions (neurophysiological, -pharmacological and -psychological) rather than of gross or microscopic neuroanatomy.

PARAMETERS OF COMPARATIVE BRAIN STRUCTURE AND FUNCTION

Numerous aspects of brain structure have evolved in parallel with the increased intellectual capacities of some species and with the specialized ecological adaptations of others. Of particular interest to neuroscientists of our own species has been the search for some emergent parameter of the human brain which sets us apart from species that seem intellectually inferior. In this regard such gross factors as increased brain size, brain/body ratio, and encephalization have long been cited, along with the relative enlargement of the cerebral cortex and its fissurization.

Inter-species differences in more fine-grained structural features have also been noted. Jerison (10) points out that evolutionary trends towards increased brain size are paralleled by increases in the proportion of glia to neurons and in the volume of dendritic arborizations, together with decreases in neuronal packing density. These factors seem to reflect a substantial enhancement of interneuronal connectivity in the cortex of large-brained species. Interspecific differences in the proportion and organization of neuropile and in the richness of cortico-cortico fiber connections have also been observed (2,13,25). Although there are definite evolutionary trends in these anatomical features, the human brain does not seem to embody any radical departures at the gross or microscopic level that have been linked with specific mental states or intellectual capabilities.

Welker (25) has catalogued dozens of neurophysiological phenomena at different levels that are candidates for showing evolutionary trends in parallel with behavioral evolution. These include mechanisms of spike conduction, synaptic interactions and integrations, glial/neuronal relationships, small circuit transactions, patterned operations of larger neuronal systems, etc. Again, however, it is not clear which, if any, single parameter of neurophysiological activity is critical for higher mental functioning.

It seems more likely that inter-species differences in mental functions should arise from the more complex aspects of brain circuitry - from the ways in which the basic neural elements are functionally interconnected - rather than from differences in those elements themselves. In the present paper we survey several diverse aspects of brain organization and function, some of which appear to be associated with reasonably specific mental functions in man. These include: 1) the organization of neurotransmitter systems, 2) electrophysiological response patterns, 3) lateral specialization of the cerebral hemispheres, and 4) functions of the neocortical association areas.

By comparing these functions in man with those of animals at various phylogenetic levels, we may be in a position to make some inferences about the mental capabilities of different species.

NEUROTRANSMITTER SYSTEMS

Based on present evidence, it is possible to delineate at least three major, chemically distinctive classes of cerebral neurotransmitters and to specify the general nature of their molecular actions. These three classes, all of which have been shown to exist in human cortex, are the amino acids (GABA, glycine, glutamate, aspartate), the monoamines (noradrenaline, dopamine, serotonin, acetylcholine), and several peptides. On the basis of current evidence, it appears that there are only minor variations in the ratios of these transmitters across species. In experiments where relatively large pieces of neocortex were homogenized and extracted, the same general crop of neurotransmitters has been seen in a variety of species, with nothing unique in either the quantity of any given transmitter or in its ratio to other transmitters to distinguish the human or primate cortex from those of other mammals. However, there does seem to be a relative enhancement of the dopamine (DA) to norepinephrine (NE) ratio in the frontal cortical fields of the rhesus monkey (Brown and Goldman, see (1)) and some enhanced content of the neuropeptides Substance P, Vasoactive Intestinal Peptide, and somatostatin in primate cortex over rodent (Emson and Hunt, see (22)).

A critical issue to be addressed in future work is the degree to which specific patterns of synaptic innervation and changes in the complexity of cortical columnar organization may represent critical structural units which underlie the operation of higher cortical functions (22). The cortical columns are composed of groups of vertically-oriented cylindrical arrays and consist of ensembles of hundreds to thousands of neurons. For cognitive phenomena, it may be necessary for specific columns to be functionally associated with other columnar groups in adjacent or distant regions of cortex, to which they can be rapidly and reversibly linked by intracortical and extracortical afferent systems. The extracortical monoaminergic afferent projections have cytological distributions which are organizationally well disposed for such linkage functions (18). It could be reasoned that the similar dimensions of cortical columns found in species of varying cognitive capacities would favor the view that increased mental sophistication is based more on the absolute number of columns and their potentials for associative interactions. Given this possibility, a more functional focus would be to consider the relative cellular effects of transmitter chemicals within a given cortical ensemble and to envision how neurotransmitter systems can relate to each other in the operation of cortical information processing.

The monoamine projections to cortex are all in the "nonspecific" afferent class morphologically, arising from pontine or mesencephalic nuclei (16,17). However, in experimental animals these systems retain considerable organizational specificity within cortical layers and regions. In infrahuman primates, the dorsolateral frontal lobes are relatively rich in DA terminals, whereas the serotonin (5-HT) content is low. Brozoski et al. (1) have found that depletion of DA in the prefrontal cortex greatly impairs a monkey's ability to perform a spatial delayed alternation task, whereas comparable depletions of NE or 5-HT failed to have an effect. This suggests that the dopaminergic system in frontal cortex may participate in specific mnemonic functions in primates.

The NE fibers from the locus coeruleus project more generally throughout the cortex as well as to the sensory nuclei of the thalamus which relay information to the cortex. Disruption of these afferents in the visual cortex by 6-OHDA has been reported to delay the development of the specialized binocular properties of visual neurons under environmental influences (12). This evidence is consistent with the more general hypothesis that the release of NE by fibers projecting diffusely from the brain stem is important in the formation or consolidation of stable changes associated with learning and memory (14).

The monoaminergic afferent systems to cortex operate slowly (100's of msec), and their discharge rates in behaving rats and monkeys increase in direct proportion to the level of behavior arousal (5). The physiological effect of activity in the locus coeruleus seems to be a potentiation of <u>both</u> excitatory and inhibitory influences on target neurons in the cortex - that is, the recipient neurons are made more sensitive to other afferent inputs. In this way, the action of the coeruleo-cortical projections may enhance the signal to noise ratios of cortical target cells and, in computer terminology, "enable" specific sensory information to influence critical targets (Bloom, see (22)). The same sort of action would suppress the discharge of those target cells not receiving sensory information during that epoch. The time course of the phasic activation of this "enabling" system (e.g., by a novel sensory stimulus) is similar to that of slow event-related potentials in man (see below).

There are suggestive data that the locus coeruleus and substantia nigra (the nuclei giving rise to cortical NE and DA fibers, respectively) are somewhat more prominent in primate and human brains than in rodents, but precise cell counts are difficult to come by and detailed estimations of innervation ratios for these highly branched, fine axon systems require much more work. Despite their extensive and overlapping innervation patterns, significant topographical specificity within these systems has been demonstrated by recent methods. The 5-HT projection to neocortex has been studied in detail only for the rodent brain and appears to be even more diffuse in its cortical distribution than the NE system, although some lateralization of these projections has been reported. A significant unilateral enrichment (on the left side) of NE projections to human thalamic nuclei might represent an aspect of hemispheric specialization.

Amino acid-mediated transmission processes are illustrated by the presumptive glutamatergic afferents to the motor cortex. These excitatory systems and complementary GABA-mediated intracortical inhibitory interneurons operate on ion conductance mechanisms which rapidly (1-10 msec) and potently depolarize and hyperpolarize, respectively. Pettigrew (20) has reported that GABA-mediated cortical inhibition plays an important role in giving visual neurons their higher-order receptive field properties.

The third class of chemically-defined cortical neurons, which use peptides as neurotransmitters, all fit the morphology of intra-columnar interneurons, but neither the quality nor the specifics of their actions have been determined in any primate. Some evidence in other brain regions suggests that the peptide systems may modify the flow of information to convergent target cells by means of a nonclassical mode of transmitter action which can be viewed as complementary to the "enabling" concept hypothetically attributed to the monoamine systems. In the case of the limited number of existing tests on peptidergic systems, this mode might be termed "disenabling" since the action of several different peptides is to prevent target cells from responding to inputs mediated by other classical transmitter systems. Thus, the peptide enkephalin seems to act to inhibit the activity of neurons, primarily by disenabling the neurons from responding to excitatory transmitters such as glutamate.

DRUGS AND BEHAVIOR

One line of experimentation which can potentially relate mental processes in man to molecular, cellular and behavioral processes in animals is on the action of drugs. Such studies have been carried out extensively over the past 25 years, generally attempting to determine the mechanisms of action of psychotherapeutic drugs as a means of understanding the neural basis of the psychopathologic conditions treated by those drugs. Each of the major classes of psychotherapeutic drugs has been found to have qualitatively similar effects upon neurotransmitter chemistry and function in human and nonhuman brains and upon associated behavioral manifestations. This similarity of cellular and behavioral action strongly supports the concept that common drug-responsive elements do exist in human and nonhuman brains. To the extent that these drug-sensitive systems can be shown to mediate specific mental activities in man, an experimental approach will be available for inferring similar activities in animals.

Despite much experimental evidence in animals, considerable effort will be required to confirm the chemical, anatomical, and functional details of neurotransmitter systems operating within human cortex. If transmitter systems with unique properties are required to explain the information processing capacity of the human cortex, such properties are more likely to involve increases in the sheer numbers of neurons and their organizational complexity rather than any novel and as yet undiscovered chemical principle. In fact, all of the neurochemical systems encountered so far in mammalian cortex generally, and all of the general features of their physiological mechanisms of action, seem to be quite similar to those that have been studied in invertebrate nervous systems.

ELECTROPHYSIOLOGICAL CORRELATES OF MENTAL ACTIVITY

Comparisons of electrophysiological response patterns in animal and human brains provide another source of data for evaluating similarities and differences in their mental activities. Among the neuroelectric patterns which can be readily recorded from the human scalp and the brains of animals are the ongoing electroencephalogram (EEG), which varies with the sleep/waking cycle and other global state changes, and the evoked or event-related potentials (ERPs). Both the EEG and ERP waves represent the fluctuating electric fields that arise from the synchronous and patterned activity of large neuronal populations as information is processed therein.

EEG Waves

Ongoing EEG waves may be recorded at the surface and in the depths of primitive and sophisticated brains alike. Bullock (2) has observed that the properties of the EEG are remarkably similar in a wide variety of vertebrate species, irrespective of size, having a preponderance of energy at relatively low frequencies (5-35 Hz) and smooth, rhythmic oscillations. This contrasts with the EEGs of most invertebrates which are more "spiky" and irregular and contain significant energies above 100 Hz.

A number of vertebrate families, including most birds and mammals, display a low-voltage, high frequency EEG pattern during periods of alertness or arousal, and a higher voltage, slower EEG during periods of drowsiness and sleep (20). Moreover, when a startling stimulus is presented to a drowsy or inattentive animal, the slow wave pattern is immediately "desynchronized" to a faster EEG. This EEG alerting response is mediated by the reticular activating system and is usually interpreted as a shift from neural synchrony to the differentiated firing patterns associated with processing the unexpected inputs.

An EEG feature that is shared by canines and primates (including man) is a prominent band of rhythmic activity in the alpha band (8-13 Hz). The increased tendency for the EEG to show a peaked "resonance" at the alpha frequency has been linked to an increase in the number of cortico-cortical connections (12). Alpha power is generally increased during states of relaxed wakefulness and is desynchronized by stimuli which provoke visual attention. In man, the relative alpha power between the two cerebral hemispheres reportedly varies as a function of verbal versus nonverbal thinking, with the presumably more active hemisphere showing less alpha activity (i.e., greater desynchronization). The EEG manifestations of the sleep/waking cycle are common to the majority of warm-blooded vertebrates (11). The cyclical pattern of slow wave sleep alternating with periods of REM sleep is found in primates, carnivores, ruminants, and birds. Assuming that REM sleep is associated with "dreaming" in lower animals as well as ourselves, Jouvet has speculated that this brain state is important for the programming of instinctive behaviors under the influence of genetic and epigenetic factors; possibly this involves a reprogramming of synapses in the forebrain, by means of the ascending monoaminergic projections.

The similarities of the basic EEG patterns among many mammalian and avian species, particularly those associated with states of conscious awareness in man (alpha rhythm, desynchronized EEG, REM sleep pattern), suggest that processes resembling consciousness have a wide phylogenetic distribution. A further implication is that the general principles of information handling and coding in the brain may not be radically dissimilar among warm-blooded animals.

EVENT-RELATED POTENTIALS (ERPs)

Superimposed upon the ongoing EEG and ERP waves that are time-locked to specific sensory, motor, and/or psychological events. For example, a sensory stimulus elicits a series of positive and negative ERP waves (also known as evoked potentials) which arise from the progressive activation of the sensory pathways and cortical sensory areas. These sensory ERPs may be readily recorded from the scalp in man using computerized signal averaging techniques or directly from the brains of experimental animals (3).

The ERP components evoked within the first 50 msec post-stimulus are quite stable and independent of the state of the subject. In contrast, the late cortical waves (later than 100 msec) show considerable variation both as a function of the state of the subject and of the specific way in which stimuli are being processed (i.e., their task-relevance).

Various types of these late ERPs in man have been associated with processes of perception, attention, recognition, expectation, and decision making (3). In the present context, we would like to point out some possible homologies between animal and human ERPs.

One class of ERP that appears to be virtually identical in man and monkeys includes the slow preparatory potentials, the CNV or "expectancy wave" and the motor readiness potentials. The CNV is a slow negative ERP which arises at the surface of the brain during periods of sensory expectation or motor preparation. Studies in monkeys have found that the CNV is generated in the superficial cortical layers and is paralleled by slow potential changes in deeper brain structures. The similarities of the CNVs in different primate species suggest a communality of brain mechanisms for these preparatory processes. CNV-like slow potentials have also been observed in cats and rats.

Strong similarities in temporal patterning have also been observed between ERP components that are associated with visual selective attention in man and neuronal discharges of the attentive monkey's parietal lobe. When a person focuses attention on light flashes in one visual field and ignores flashes in the opposite field, the attended flashes elicit an enhanced negative ERP over the parietal scalp, beginning at about 100 msec (Fig. 1). Recordings of single neurons from the rhesus parietal lobe (see (27)) show a phasic burst of activity which has a similar time course (allowing for the smaller brain size) to the human ERP and is similarly incremented when flashes are made task-relevant. These electrophysiological parallels, together with the similarities of cortical lesion effects (see below), suggest that monkeys and man utilize similar mechanisms for directing and focussing their visual attention. ERP signs of selective attention have also been demonstrated to auditory and somatic stimuli, and this technique should allow further comparisons between man and animals in mechanisms of attention and perception.

The most thoroughly studied ERP in man is the P3 or P300 wave, a positive wave at around 300 msec latency. The P300 is elicited by stimulus events that are relevant and attended but occur unexpectedly. For example, if a person is required to detect an occasional target stimulus which occurs unpredictably among a sequence of non-targets, the target will elicit a P300. P300 waves are also elicited by novel, surprising stimuli, probably as a component of the orienting response.

Several proposals have been advanced as to the psychological correlates of the P300. Early hypotheses linked the P300 to information delivery, orientation, and cognitive evaluation. More recently, Donchin et al. (see (3)) have emphasized the dependency of P300 upon cognitive structures in memory; they suggest that P300 is a correlate of the "updating" of neuronal models of the environment following unexpected stimuli. Since the P300 seems to be elicited primarily (if not exclusively) when a person is conscious of a stimulus event, this ERP has considerable promise as a marker of conscious processing in other species.

Several recent experiments have sought to determine whether animals have P300-like ERPs to unexpected stimuli. In squirrel monkeys, Neville and Foote (in preparation) have found that unexpected deviant sounds in a sequence of tones (either a frequency shift or a novel "bark" sound) are followed by an ERP that closely resembles the P300 elicited by the same stimuli in human subjects. Studies in cats have also found positive ERPs in the 200-300 msec range following unexpected stimuli with conditioned significance (26). Further studies are needed, however, to establish whether these candidate-P300's in animals have the functional properties of the P300 in man and whether their neuronal generator systems are homologous.

Comparisons of other types of ERPs across species may provide insights into different cognitive functions. For instance,

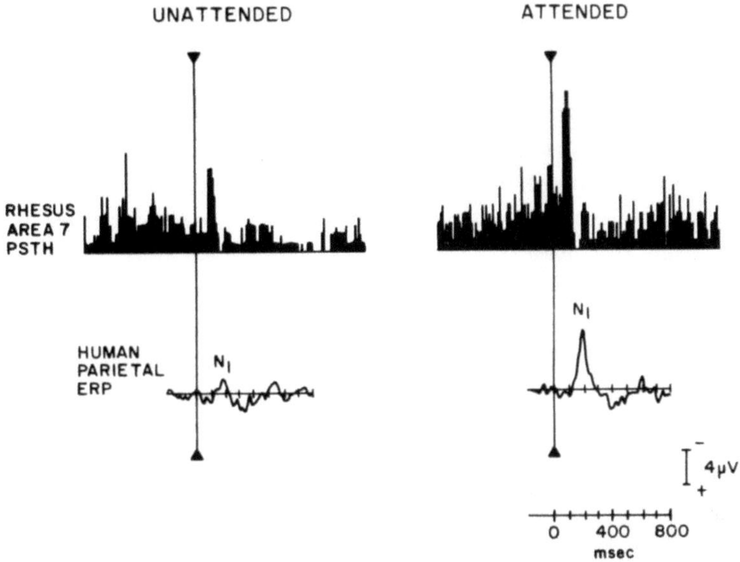

FIG. 1 - Similarity of attention effects on single unit discharge in rhesus parietal lobe and human parietal ERP. Rhesus data are post-stimulus time histograms (PSTH) of the activity of a single neuron in area 7, recorded with a microelectrode. This plot represents the average number of neuronal spike discharges in successive time intervals (bins) before and after the stimulus, the onset of which is designated by the vertical line. The human data are computer-averaged ERPs recorded from the scalp over several hundred stimulus presentations.

Kutas and Hillyard (15) have reported a late negative ERP (N400) following incongruous words in sentences, which they hypothesize may be a specific marker of semantic analyses during language processing. If this turns out to be the case, the identification of an N400 homolog in other species would carry important implications for their mental functioning.

LATERALIZATION OF FUNCTION

A very salient feature of the human brain is the specialization of the left and right cerebral hemispheres for different linguistic, cognitive, and perceptual functions. The left hemisphere (in most right-handed individuals) is specialized for language, arithmetic, and logical thought, while the right hemisphere excels at visual perception, particularly of spatial relations and other non-language skills (see Levy, this volume). Some authors have proposed that cerebral laterality occurs in direct proportion to the evolution of language, of handedness, of executive control, or the existence of a conscious self. The central question we would like to consider here is whether the human faculties for language, logical thought, consciousness, etc., are uniquely related to and dependent upon this pattern of specialization. To this end, it is important to consider the evidence for comparable hemispheric specialization in other vertebrates (4,24) to see whether the human pattern of cerebral laterality sets man sharply apart from lower primates and other animals.

Cerebral specialization in man has definite neuroanatomical correlates. The planum temporale, an auditory association area of the temporal lobe, is larger in the left hemisphere in most brains, and the left Sylvian fissure is longer. There are also tendencies for the occipital pole to be larger on the left, whereas the frontal lobe is wider on the right, asymmetries that vary with a person's handedness (6). Left-right asymmetries less marked than those of man have been reported in the temporal lobes of the great apes (chimp, gorilla, and orangutan) but were not seen consistently in Rhesus or New World monkeys. Thus, there may be a progression in the degree of asymmetry in temporal brain regions from man to apes to monkeys. Lateral asymmetries have been studied less extensively in other species, but it is worth noting that rats may show a thicker cortex mantle in the right hemisphere, and songbirds appear to have no obvious anatomical brain asymmetries despite a marked functional lateralization.

Functional Asymmetries in Nonhuman Mammals

There does not appear to be any overall tendency towards right- or left-handedness in monkeys, although strong individual preferences may be seen (7). The same general picture seems true for rats, mice, cats, and possums, with little data available for apes. Some modest evidence for functional specialization in several Rhesus monkeys was obtained by Dewson, who found that left-sided lesions impaired sequential auditory match-to-sample task more than did right-hemisphere lesions. However, a number of other types of learning tasks have not revealed any lateralized preferences, either in split-brain monkeys or following unilateral lesions. More recently, Vermeire and Hamilton (23) reported that split-brain monkeys were better at making sequential same-different judgements using the hand contralateral to their preferred hand, while the ipsilateral hemisphere showed stronger visual preferences in a free observing situation. They suggest that these correlations of hemisphericity with handedness might parallel those seen in rats and man, and that a general basis for hemispheric specialization might be quantitative asymmetries in emotional and/or attentional processes.

Peterson et al. (see (24)) have reported that Japanese macaques are better able to discriminate their species-specific calls with the right ear than the left ear. This resembles the right ear (left hemisphere) preference for discriminating speech sounds seen in man under certain conditions, leading the authors to conclude that these macaques "engage left hemisphere processors for the analysis of communicatively significant sounds that are analogous to the lateralized mechanisms used by humans listening to speech." Their data suggest that hemispheric lateralization may arise phylogenetically in association with an increasingly complex communicative repertoire.

Denenberg (4) has proposed that asymmetry is an initial condition of vertebrate brains, with the left hemisphere predisposed for receiving and transmitting communications and the right specialized for dealing with "spatial and affective matters." This lateralization develops ontogenetically as a result of early experience and competition between homologous brain areas. If it can indeed be shown that functional brain asymmetry has such a general basis, then the degree of lateralization in a species may well become an important criterion for assessing its mental qualities and capabilities. At present, however, the nature and origins of lateral specialization are not understood well enough to reach firm conclusions along these lines.

FUNCTIONAL ORGANIZATION OF THE CEREBRAL CORTEX

The evolutionary trend towards enlargement of the cortical association areas in higher primates has often been related to increasing sophistication of sensory processing and cognitive capability. There is now sufficient evidence available for comparisons to be made between man and other primates as to the functional organization of the cortical association areas, again looking at the question of whether any unique areas or functions have evolved in man.

Horel (8) has reviewed studies of temporal lobe lesions in monkeys and man and has concluded that severe learning disorders are the major sequelae in both species. Comparing the behavioral performance of human amnesics with temporal-lesioned monkeys, Horel concluded that both display a similar syndrome that includes retrograde and prograde amnesia. Iversen and Weiskrantz (9) have postulated an evolutionary homology between the monkey's inferotemporal cortex and the human ventral-medial temporal lobe, which play homologous roles in learning, memory, and visual discrimination processes. Thus, there is good evidence that very similar perceptual and mnemonic functions are subserved by temporal lobe structures in man and monkeys.

It has been suggested that the human inferior parietal lobule (particularly the angular and supramarginal gyri) may have unique functions as an intermodal "association area of association areas" that is absent in monkeys and poorly developed in apes. However, neuroanatomical studies by Pandya and Kuypers (19) have found areas in the Rhesus inferior parietal lobe where polymodal projections do converge. Moreover, unilateral lesions of the inferior parietal area produce similar behavioral syndromes in monkeys and humans, characterized by a neglect of contralateral stimulation, without a primary sensory deficit. On the basis of lesion and electrophysiological evidence, Mountcastle (see (27)) has proposed that the parietal lobes of both man and monkeys contain the essential neural mechanisms for orientation and exploration within extrapersonal space. These mechanisms include the functions of directed visual attention.

Frontal lobe functions in man and monkeys also show a number of similarities. In monkeys, lesioning the prefrontal association cortex produces the well-known impairment of delayed response performance, as well as perseverative deficits that are manifest in reversal learning, learning sets, and other tasks (Goldman et al., see (1)). In man, the frontal lobe syndrome is generally characterized by a tendency towards response perseveration and difficulty in shifting strategies or sets, often accompanied by a poor attention span, distractability, and social and motivational peculiarities. In man, however, the delayed response deficit does not seem as prominent as in monkeys. Possibly, these frontal lobe syndromes might have common elements such as a deficiency in the focussing and sustaining of attention on a task.

From this brief survey it is evident that the major association areas of the cortex seem to function in a similar fashion in man and in other higher primate species. Since these cerebral regions in man are believed to mediate some of the most sophisticated of mental acts, it seems reasonable to conclude that their evolutionary homologs in primates engage in similarly complex forms of processing, probably associated with at least qualitatively similar levels of conscious awareness.

REFERENCES

(1) Brozoski, T.J.; Brown, R.M.; Rosvold, H.E.; and Goldman, P.S. 1979. Cognitive deficit caused by regional depletion of dopamine in prefrontal cortex of Rhesus monkey. Science 205: 929-932.

(2) Bullock, T.H. 1974. Comparisons between vertebrates and invertebrates in nervous organization. In The Neurosciences Third Study Program, eds. F.O. Schmitt et al., pp. 343-346. Cambridge: MIT Press.

(3) Callaway, E.; Tueting, P.; and Koslow, S., eds. 1978. Event-Related Brain Potentials in Man. New York: Academic Press.

(4) Denenberg, V.H. 1981. Hemispheric laterality in animals and the effects of early experience. Behav. Brain Sci. 4: in press.

(5) Foote, S.L.; Aston-Jones, G.; and Bloom, F.E. 1980. Impulse activity of locus coeruleus neurons in awake rats and monkeys is a function of sensory stimulation and arousal. Proc. Natl. Acad. Sci. 77: 3033-3037.

(6) Galaburda, A.M.; LeMay, M.; Kemper, T.L.; and Geschwind, N. 1978. Right-left asymmetries in the brain. Structural differences between the hemispheres may underlie cerebral dominance. Science 199: 852-856.

(7) Hamilton, C.R. 1977. An assessment of hemispheric specialization in monkeys. Ann. New York Acad. Sci. 229: 222-232.

(8) Horel, J.A. 1978. The neuroanatomy of amnesia: A critique of the hippocampal memory hypothesis. Brain 101: 403-445.

(9) Iversen, S., and Weiskrantz, L. 1964. Temporal lobe lesions and memory in the monkey. Nature (London) 201: 740-742.

(10) Jerison, H.J. 1979. The evolution of diversity in brain size. In Development and Evolution of Brain Size, eds. M.E. Hahn et al., pp. 29-57. New York: Academic Press.

(11) Jouvet, M. 1975. The function of dreaming: A neurophysiologist's point of view. In Handbook of Psychobiology, eds. M.S. Gazzaniga and C. Blakemore, pp. 499-527. New York: Academic Press.

(12) Kasamatsu, T., and Pettigrew, J.D. 1976. Depletion of brain catecholamines: Failure of ocular dominance shift after monocular occlusion in kittens. Science 194: 206-208.

(13) Katznelson, R.D. 1981. Normal modes of the brain: neuroanatomical basis and a physiological theoretical model. In Electric Fields of the Brain, ed. P.L. Nunez. Oxford University Press.

(14) Kety, S.S. 1970. The biogenic amines in the central nervous system: Their possible roles in arousal, emotion, and learning. In The Neurosciences Second Study Program, ed. F.O. Schmitt. New York: Rockefeller University Press.

(15) Kutas, M., and Hillyard, S.A. 1980. Reading senseless sentences: Brain potentials reflect semantic incongruity. Science 207: 203-205.

(16) Moore, R.Y., and Bloom, F.E. 1978. Central catecholamine neuron systems: Anatomy and physiology of the dopamine systems. Ann. Rev. Neurosci. 1: 129-169.

(17) Moore, R.Y., and Bloom, F.E. 1979. Central catecholamine neuron systems. Ann. Rev. Neurosci. 2: 113-168.

(18) Morrison, J.H.; Molliver, M.E.; Grzanna, R.; and Coyle, J.T. The intra-cortical trajectory of the coeruleocortical projection in the rat: A tangentially organized cortical afferent. Neuroscience 6: 139-158.

(19) Pandya, D.N., and Kuypers, H.G.J.M. 1969. Corticocortical connections in the rhesus monkey. Brain Res. 13: 13-36.

(20) Pettigrew, J.D., and Daniels, J.D. 1973. Gamma-aminobutyric acid antagonism in visual cortex: Different effects on simple, complex, and hypercomplex neurons. Science 182: 81-83.

(21) Pfurtscheller, G.; Buser, P.; Lopes da Silva, F.H.; and Petsche, H., eds. 1980. In Rhythmic EEG Activities and Cortical Functioning. Developments in Neuroscience 10. New York: Elsevier/North Holland.

(22) Schmitt, F.O.; Worden, F.G.; Adelman, G.; and Dennis, S.G., eds. 1981. The Organization of the Cerebral Cortex. Proceedings of a Neurosciences Research Program Colloquium, in press.

(23) Vermeire, B.A., and Hamilton, C.R. 1981. Hemispheric specialization in split-brain monkeys. Soc. for Neurosci. Abst. 6: 811.

(24) Walker, S.F. 1980. Lateralization of functions in the vertebrate brain: A review. Brit. J. Psychol. 71: 329-367.

(25) Welker, W. 1976. Brain evolution in mammals. In Evolution of Brain and Behavior in Vertebrates, eds. R.B. Masterton et al., pp. 251-344. Hillsdale, NJ: Erlbaum.

(26) Wilder, M.B.; Farley, G.R.; and Starr, A. 1981. Endogenous late positive component of the evoked potential in cats corresponding to P300 in humans. Science 211: 605-607.

(27) Wurtz, R.H.; Goldberg, M.E.; and Robinson, D.L. 1980. Behavioral modulation of visual responses in the monkey. In Progress in Psychobiology and Physiological Psychology, eds. J.M. Sprague and A.N. Epstein, vol. 9, pp. 43-83. New York: Academic Press.

Animal Mind - Human Mind, ed. D.R. Griffin, pp. 33-56.
Dahlem Konferenzen 1982. Berlin, Heidelberg, New York: Springer-Verlag.

Some Perspectives on the Evolution of Intelligence and the Brain

W. Hodos
Dept. of Psychology, University of Maryland
College Park, MD 20742, USA

Abstract. The twin searches for animal intelligence and its neural counterparts have been dominated by a model that predicted a unilinear, hierarchical progression from simpler intellectual abilities to more complex abilities and from simpler brains to more complex brains. This model has produced a number of baffling inconsistencies and paradoxes for investigators who sought a smooth progression of increased behavioral performance or increased brain complexity among animals selected to represent stages in human evolution. On the other hand, a model based on the evolutionary principles of divergence and adaptation can deal more readily with the nonlinearities observed in both the behavioral data and the neuroanatomical data.

THE NATURE OF ANIMAL INTELLIGENCE

When we ask the question, "Is a human more intelligent than a cat?", the answer seems so obvious as to hardly merit a reply. But when we ask, "Is a cat more intelligent than a dog?", the answer is not at all obvious. Moreover, the respondents to this latter question would likely fall into three groups: (a) those who dislike animals in general (who we dismiss from further consideration), (b) cat lovers, who will ardently extol the intellectual prowess of the objects of their affection, and (c) dog lovers, who will sing the praises of the acumen and perspicacity of man's and woman's best friend. The differences in the answers to these two questions illustrate an important problem in the assessment of the intelligence of any animal; namely, that intelligence is not a biological property of the organism as is height, weight, hormone level, calcium concentration, etc. The decision as to whether a particular behavior is intelligent or not is a value judgment on the part of the observer, like personal beauty, artistic beauty, honor, etc. Moreover, behavior is seen as intelligent or not according to the context in which it occurs. The same pattern of responses that in one situation is regarded as highly intelligent and adaptive might be regarded as very unintelligent and self-defeating in another situation.

Behaviors reminiscent of those that are of value to contemporary western, human society are those that are regarded as intelligent when found in animals by contemporary western scientists. The absence of such human-valued behaviors often is taken as a sign of "stupidity." A further difficulty is the lack of general agreement about which are the appropriate behaviors to measure. In other words, even within the class of human-valued behaviors, there is not universal agreement about which behaviors should be regarded as indicative of intellectual capacity.

In the brief space allotted to me, I shall discuss the nature of animal intelligence and its relation to human intelligence and offer some suggestions about directions for future study. I shall then discuss some of the changing ideas about the evolution of the brain, especially those regions that traditionally have been regarded as being of particular importance for the evolution of intelligence. This paper is not intended as a comprehensive review of the literature in these fields, but rather an attempt to acquaint the reader with some general principles and theoretical positions. References to comprehensive review articles are given at appropriate points for those who require more detailed information.

Early Ideas About the Nature of Intelligence

A serious interest in animal intelligence predates work on the formal assessment of human intelligence. Romanes' publication of Animal Intelligence in 1882 (26) marks the first attempt at a scientific analysis of animal intelligence. Romanes defined intelligence as the capacity to adjust behavior in accordance with changing conditions. He used phrases such as "intentional adaptation" and "conscious choice" to differentiate the more intelligent from the less intelligent of animals. On the other hand, Lloyd Morgan (17), a contemporary of Romanes, discussed intelligence in terms of the "psychical level" of an animal. His opinion was that one must assess the animal mind in the context of the only mind of which we have first-hand knowledge; i.e., the human mind. Lloyd Morgan's book, An Introduction to Comparative Psychology (17), which was published in 1894, postulated the existence of a "psychological scale" based on a hierarchy of faculties. This was an era when faculty psychology was still an influential doctrine. Faculty psychology enjoyed considerable popularity throughout the 19th and early 20th centuries. Among the faculties (i.e., powers of the mind) were perception, judgment, memory, self-preservation, pity, morality, self-esteem, etc. Faculty psychology was the basis for Gall's efforts in attempting to localize function (faculties) in the cerebral hemisphere (4). Lloyd Morgan's position was that higher faculties evolved from lower faculties. The presumed existence of this unilinear, progressive scale of behavioral development was a natural companion to the presumed unilinear, progressive scale of morphological development that was the popular notion of the course of evolutionary history. This doctrine has influenced the comparative study of animal intelligence until the present day.

The first standardized test of human intelligence was published by Binet in 1905 (27). Although Binet's early ideas about intelligence grew out of faculty psychology, his 1905 intelligence scale represented a radical departure from the measurement of faculties to the measurement of intelligence in general; i.e., an assessment of overall intellectual aptitude rather than measurement of individual faculties. This latter step was taken at about the same time by Spearman who proposed the concept of "general intelligence", which is a single basic factor that manifests itself in the person's performance on tests of specific intellectual abilities. Although it has undergone considerable refinement and sophistication by the application of the statistical techniques of factor analysis, the basic idea of Spearman, of a general intellectual factor that expresses itself in a wide variety of intellectual activities, remains in vogue today. A history and analysis of changing views about the nature of human intelligence has been provided by Tuddenham (27).

Contemporary Views on the Nature of Intelligence

How can we characterize general intelligence so that we are able to recognize it in both human and animal behavior? Some comments that Tuddenham (27) offered at the end of his historical survey may be helpful in this regard: "...intelligence is not an entity, nor even a dimension _in_ a person, but rather an _evaluation_ of a behavior sequence (or the average of many such), from the point of view of its adaptive adequacy. What constitutes intelligence depends upon what the situation demands..." [p. 517]. The idea of human intelligence as an abstract characterization of an organism's behavioral responses to pressures from the environment appears to be the prevailing view among contemporary theorists. The notion also seems well suited to animal intelligence and is remarkably close to Romanes' view. It is a very different formulation from that of Lloyd Morgan's hierarchy of faculties with the higher faculties emerging from the lower.

The topic of intelligence is mentioned only in passing, if at all, in contemporary textbooks of comparative psychology or animal behavior. When intelligence is mentioned, it is not defined but merely given as if its meaning were universally understood. Such examples that are offered to demonstrate that animals are capable of "higher mental processes" or "complex behavior" are nearly always studies of insight, reasoning, problem solving, concept formation, matching-to-sample, reversal learning, learning set, tool use, etc; i.e., behaviors that are of value to humans. The conception of intelligence as a general adaptive behavioral response to the pressures of the environment rather than as specific intellectual faculties seems not to have found its way into the realm of animal research to any great extent. Moreover, Lloyd Morgan's ideas about higher faculties being derived from lower faculties in a unilinear fashion seems still to be with us. Two notable exceptions to this approach may be found in the work of Bitterman (3), whose efforts at the comparative analysis of learning have been directed towards the search for divergence and nonlinearity in learning mechanisms, and an insightful article by Riopelle and Hill (25) on complex behavior in animals.

To view animal intelligence as being merely a scaled down version of human intelligence errs in two ways: (a) it ignores the special nature of human intelligence, and (b) it offers an excessively narrow conception of the nature of animal intelligence. Human intelligence is special because of language and related cognitive skills, which permit us to communicate not only with each other, but with past and future generations. Each generation of humans benefits from the cumulative knowledge of previous generations. We can profit from the errors of the past, without actually experiencing them ourselves, by following symbolic instructions from others. Other animals (with the possible exception of great apes) seem to lack this ability. However, they have special capabilities for their own environments that suit them very well even though these abilities may be of limited value to us.

A refreshing perspective on the nature of vertebrate intelligence may be found in a discussion of the intelligence of invertebrates by Corning, Dyal and Lahue (7).

> "The imposition of vertebrate biases on invertebrates predisposes thinking that what is intelligent behavior

> for the vertebrate must be a useful and desirable capacity in the invertebrate. For example, the capacity to associate stimuli is probably one of the basic requirements for vertebrate intelligence but what is intelligent for the vertebrate need not represent intelligence for the invertebrate; in annelids 'the existence of such learning may prove to be of much more significance to the animal behaviorist than to the worm. While it may well be that worms <u>can</u> learn associatively the demands of their normal environments seldom, if ever, actually require that they do so...'." [pp. 216-217].

These authors discuss invertebrate intelligence in the context of "adaptive organism-environment transactions," which seem to be similar to the global definition of intelligence suggested by Tuddenham and quoted above.

Intelligence as a Population Parameter

Before ending my discussion of human and animal intelligence, I must bring one additional caveat to your attention. Intelligence tests since Binet's time have always been devices for comparing an individual drawn from a specific population with the other members of that population. To the extent that they have been used for this purpose, they have been relatively successful and have gained a broad acceptance. However, when testers have drawn an individual from one population and tested him or her with a test designed for a different population, the results have been less than satisfactory. Attempts at cross-cultural intelligence tests have largely been failures or had dubious success at best. The most recent manifestation of the difficulties of comparing the intelligence of different populations is the current controversy about racial differences in intelligence. However, the argument is quite germaine to the problem of the comparison of the intelligence of different species, orders or classes of vertebrates, or the comparison of vertebrate intelligence with that of invertebrates. For example, in assessing the intelligence of a frog, we want to know how well adapted is this frog to a frog's environment, not how well it responds to environmental pressures that are appropriate for a mammal.

The nature of intelligence tests, whether animal or human, is such that measurements can only be made on individuals who are samples of a population. If we wish to compare the intelligence of two populations, we must decide whether to design separate tests for the two populations to take into account those factors that make the populations different or to give both populations the same test. If we select the first option, we run the risk that the two tests may not be comparable and any differences that we observe may be due to differences in the test rather than differences in the populations. If we give both groups the same test, we run the risk that the test may favor one group more than the other and again may produce spurious differences. A difficulty that is enhanced in animal comparisons is that the more closely related the animals are (for example, the same genus), the more their variance envelopes may overlap, which could increase the difficulty of seeing differences. On the other hand, if the groups being compared differ greatly (e.g., different classes), the liklihood of nonequivalence of tests becomes greater. The problems associated with the comparison of closely related animals can be dealt with statistically and by the construction of tests with high resolving power. However, neither of these procedures will help with the problem of test inequality. Therefore, comparisons of the intelligence of closely related animal populations will probably be more meaningful than comparisons of animals with more remote common ancestry.

EVOLUTIONARY TRENDS IN THE INTELLIGENT BEHAVIOR OF ANIMALS

Research in animal behavior that has been directed at adducing evidence in support of Lloyd Morgan's "psychological scale" has generally been disappointing because the expected smooth progression of abilities did not emerge. Certain abilities, such as learning, seemed to be very well developed in so many taxonomic groups that they came to be regarded as useless for differentiating species according to their degree of presumed relatedness to humans. The only types of "intellectual" skills that seemed to produce the desired hierarchy were what could loosely be called complex behavior such as rule learning, abstraction, tool use and the like. Even so, a number of disquieting puzzles remained: Why should some birds perform better than some mammals, including some primates? Why should some nonprimate mammals perform better than some primates? These questions remain puzzling only if one accepts the "psychological scale" as a model for the evolution of intelligence. However, if one accepts the idea of divergence in the evolution of intelligence rather than a smooth progression of abilities, then these seemingly anomalous observations become quite plausible.

Some Principles of Evolution and Adaptation

A principal culprit in the dissemination of false leads in the study of animal intelligence has been the notion that evolutionary trends follow a unilinear progression from simpler to more complex. This idea has been seductive for two reasons. First, it is elegantly simple. Second, it is sufficiently similar to some actual evolutionary trends that casual, intermittent contact with the evolution literature would seem to confirm its validity. However, the historical record of life on this planet, whether animal or plant, vertebrate or invertebrate, indicates that those organisms that exist today are here not because of an advance along a single road with ever increasing complexity, but because of divergence along a multiplicity of roads and by adaptation to whatever environment through which those diverging roads led. In some cases, that adaptation called for increasing complexity. In other cases, it called for increasing simplicity. When some roads led to similar environments, the "travellers" often adapted in similar ways which resulted in the appearance of similarity between organisms that did not share an immediate common ancestry. So striking were some of these similarities that they led to incorrect conclusions about the relatedness of the animals. This problem has been discussed in greater detail elsewhere (6, 12, 13, 20).

The implications of the principles of divergence and adaptation for understanding the evolution of the brain and intelligence or any other behavioral manifestation are that we must not expect to find a smooth, steady progression from group to group as we attempt to reconstruct history from the study of living animals. We must be prepared to find nonlinearities in the developmental history of a lineage. We must be prepared to find structures and mechanisms waxing and waning over and over as animals leave one environmental niche and enter another. We must be prepared to find equivalent levels of structural or behavioral sophistication emerging along different branches of the evolutionary tree as animals that may be only remotely related adapt to similar pressures from the environment.

Another point that must be considered by students of evolutionary progress is that evolution does not progress at the same rate in all systems of a given lineage (21). Thus an animal may have achieved a high level of efficiency or specialization in some regions of its brain or in some aspects of its behavior and retained a relatively unsophisticated level of organization or activity in another. Therefore, one can be grossly misled by characterizing an animal as being "primitive" or "advanced" outside of a specific context or apart from a specific criterion (11). Thus, a human being would be rated as being at a higher grade of development than a shark on the basis of its ability to see fine detail and to solve abstract problems, but at a lower grade on its ability to detect low frequency pressure waves or electric fields in the surrounding environment or in its ability as a predator in the marine environment. Moreover, the rates of evolution vary in different lineages so that animals with rather different ancestors may have achieved the same grade or level of organization in some systems but not in others.

Demonstrations of Animal "Intelligence"

Even though intelligence is rarely defined in animal studies, considerable data have been collected and offered as evidence of animal intelligence. These data span the range from simple associative learning to complex learning of rules and strategies to tool use. Although a number of these behavioral phenomena fail to satisfy a definition of intelligence based on the adaptive demands of the particular animal's environment, they are widely cited and will be reviewed briefly here so that they may be evaluated in the light of the foregoing discussion. The literature on the types of tests that have been used to assess animal intelligence is too vast to summarize in this small space. The reader will find several comprehensive review articles in the reference list (2, 8, 25, 28, 29). However, a few generalizations will provide some perspective on the current state of knowledge. For the present purposes, let me classify behaviors that have been called "intelligent" into four categories: habituation, classical conditioning, instrumental conditioning, and complex behavior. Habituation appears to be a universal phenomenon in animal organisms, including protozoans. Classical and instrumental conditioning have been demonstrated to the satisfaction of most workers in those metazoans that possess a central nervous system with axial symmetry. Included in this category are platyhelminthes (such as planarians), annelids (such as earthworms), arthropods (such as cockroaches and crabs), molluscs (such as the octopus), and all vertebrate classes.

In the category of complex behavior, we find that delayed alternation and delayed response have been reported in arthropods and in molluscs. Molluscs also are capable of reversal learning. These behavioral abilities have been demonstrated in vertebrates as well. Thus, teleosts, birds and mammals have all demonstrated the ability to improve their performance after successive reversals.

Learning sets have been the most extensively used tests to compare the intellectual abilities of animals. Successful performance of this task involves the development of a response "strategy" or "rule" of continuing to respond to a stimulus if it was rewarded on the first trial. If the first response was unrewarded, the animal must switch to the other stimulus and stay with it. This strategy permits rapid transfer from one discrimination problem to the next. An animal that does not make use of the rule shows little improvement in the rate of learning of each successive new problem even after many hundreds of such problems have been presented. Visual learning sets have been interesting because humans and great apes perform extremely well on them, rhesus monkeys less well, cats still less well, and rats seem to apply this rule sparingly if at all. However, some carnivores, such as minks and ferrets, perform as well as old world monkeys. Moreover, birds such as blue jays, mynas and pigeons have performed at levels on a par with those of primates. When rats are permitted to perform this task using olfactory stimuli rather than visual, they acquire and apply the rule with a rapidity equal to that of the great apes and humans tested with visual stimuli.

The fabrication and use of tools is an ability that until recently had been regarded as the sole province of human beings. If tool use is defined as the manipulation of an inanimate object, not internally manufactured, in order to change the location or structure of another object, then we find the use of tools to be present in rather widely separated branches of the evolutionary tree. Insects of the order Diptera have been described to capture prey by showering them with sand particles. The archer fish squirts droplets of water from its mouth, which knock down insects perched nearby. A number of species of birds have been reported to use twigs, cactus spines, and similar objects to dislodge prey from hiding places. Captive blue jays have been reported to tear strips of newspaper to rake in grain that had scattered beyond the reach of their bill. African vultures use stones to break open ostrich eggs. Sea otters hammer open shellfish with stones. Many species of primates have been reported to use twigs or branches as rakes or probes as do birds.

One could go on at great length listing examples of behaviors that have been called intelligent. They are behaviors that are valued by humans because they are viewed as being related to or components of human intelligence. However, their relevance to animal intelligence must be evaluated on the basis of the usefulness of such behavior to the animal in dealing adaptively in response to pressures in its native environment and the extent to which the response is adaptive. For example, the use of tools by animals surely represents a manipulation of the environment that is of value to the animal for the same reasons that it is of value to us. Whether this manipulation of the environment is intelligent or not depends on whether the animal is performing it in an adaptive way or merely as a reflex or pre-programmed response pattern. If the tool use represents a relatively inflexible, fixed-action pattern triggered by a narrow class of stimuli, we would not classify it as intelligent. But if the animal demonstrates that it can adapt new materials to use as tools or can use old tools to accomplish novel environmental manipulations, then the behavior would seem to fit better with the definition of intelligent behavior.

EVOLUTION OF THE BRAIN AND INTELLIGENCE

If we accept the concensus of contemporary theorists that intelligence is not an entity that can be measured directly, but an abstract conceptualization that can only be measured by its effects, how can we expect to find neural correlates of it? Jerison (14) has argued that organisms' brains have evolved in ways appropriate to their adaptations to the environment just as has any other organ system. Moreover, the greater the extent to which an animal makes use of behavior to adapt to its environment, the greater will be the number of neurons in its brain and the number of interconnections. Large numbers of neurons and interconnections ultimately will be reflected in greater volumes of the gross subdivisions of the brain. That large cerebral volumes may be due to the elaboration of different neuronal systems in two different species is of little consequence in understanding the biological bases of intelligence according to Jerison, because the two systems each have enlarged in response to selective pressures of the environment, even though these may be different in each case.

Just as the study of animal intelligence has been dominated by a search for a unilinear progression from the "lower" animals to the "higher" animals, so has the study of the evolution of the brain (6, 12, 13, 20). The notions about the origin and development of various brain systems that were formulated by the comparative neuroanatomists of the late 19th and early 20th centuries, which were based on preliminary observations and fragmentary data, assumed the status of dogma and have been repeated in successive editions of textbooks and numerous theoretical papers. Only recently have we come to realize that some of these long-standing ideas have not found support in light of modern methods of investigation. Summaries of contemporary data and ideas are given in the reference list (9, 14, 15, 18-20).

The following are some points of dogma about brain evolution that have not stood the test of time: (a) A progressive increase in brain size can be found when one compares animals in the sequence of fish, reptiles, birds, mammals. Jerison (14) has provided data that shows clearly that this is not the case. He has reported brain weight and body weight data for 198 species of vertebrates and has described large amounts of overlap between the relative weights of fish brains and reptile brains. Moreover, some fish have brains that are larger than those of reptiles of equivalent weight. Similarly, the distributions of relative brain weights of birds and mammals partially overlap. Thus, some birds have brains that would be regarded as large in mammals of equivalent body weight. Moreover, some birds have brains that weigh as much in proportion to their body weights as do the brains of primates of equivalent body weight. (b) A progressive increase in the size of the forebrain can be observed when one compares animals in the sequence lampreys, sharks, fish, amphibians, reptiles, birds, mammals. Northcutt (20) has carried out a similar analysis of forebrain weight plotted against body weights in 26 species of vertebrates. These data, although not as extensive as those of Jerison's, reveal similar relationships. They indicate that relative forebrain weight has increased in some species of every vertebrate lineage. Moreoever, birds and mammals appear to have comparable ratios of forebrain weight to body weight; some sharks have forebrains that approach those of mammals in relative size; and some bony fishes have forebrain weights that would be regarded as typical of reptiles of equivalent body weight. (c) Relative brain size is a fixed attribute of a taxonomic group throughout time. Jerison (14) has also reported that in carnivores and ungulates, a consistent trend throughout their evolutionary history has been for mean relative brain size to increase progressively through time; but the variance also increases progressively. In other words, two trends have occurred in the temporal succession of species within each of these lineages; i.e., average relative brain size has increased and diversity in brain size has increased. (d) The telencephalon of sharks and fishes is dominated by olfaction and the telencephalon evolved as a specialization of the olfactory system. The contemporary evidence indicates that the representation of the olfactory system in the telencephalon of nonmammalian vertebrates is no greater than in the telencephalon of mammals. No convincing data exist to indicate that the telencephalon of early vertebrates was wholly an olfactory organ. (e) The highly organized structures of the forebrains of "higher" animals evolved from undifferentiated primordial pools of cells in the "lower" animals. No strong evidence for the existence of these primordia has come forth. The early comparative neuroanatomists leaned heavily on data from modern amphibians as the tetrapod prototype. The relatively poorer differentiation of the forebrain structures of these animals led them to believe that this represented the primitive or primordial state. However, an examination of the forebrains of nontetrapod vertebrates indicates that their forebrain structures are quite well differentiated. The seemingly simpler organization of the amphibian forebrain thus appears to be a derived state rather than an ancestral condition. The most reasonable reading of the evidence available to date is that a well organized and differentiated central nervous system is probably an ancient property of the vertebrates and may have been derived from their invertebrate ancestors.

NEURAL CORRELATES OF INTELLIGENCE

If these traditional views of brain evolution and their presumed relationship to the evolution of intelligence have not been supported by contemporary research, are there any structural entities in the central nervous system that might prove valuable in the search for neural correlates of intelligent behavior? Recent studies have pointed out a number of interesting possibilities (10, 16, 22, 23, 24).

Spinal Cord

The spinal cord is usually one of the first regions of the central nervous system to be eliminated from consideration as a region in which to search for neural correlates of intelligent behavior. This is largely due to the fact that the mammalian spinal cord is regarded as the model of spinal cord organization and function. However, several factors should be considered before the spinal cord is deleted from the list of candidates. First, the organization of the spinal cord is rather different in tetrapods and non-tetrapods. The non-tetrapod spinal cord (and even the spinal cords of some tetrapods) have considerably more autonomy of function than do the spinal cords of familiar laboratory mammals or humans. For example, a decerebrated frog is capable of a remarkable degree of adaptive responsiveness to changes in environmental pressures. A decerebrated chicken is capable of maintaining upright posture and locomotion as well as responsiveness to some environmental demands. But even decerebrated mammals appear to be capable of some associative learning (5). To what extent these behaviors represent preprogrammed sequences or original adaptive responses remains to be seen; however, the spinal cord should not be discarded out of hand as a region in which to search for correlates of intelligent behavior.

Cranial Nerves

The cranial nerve nuclei of the medulla and pons are also usually dismissed from consideration because of their role preprogrammed reflexes. However, in many cases this assumption may be gratuitous. Many vertebrates have sensory cranial nerve nuclei that are developed to an extent far beyond anything found in the comparable cell populations in humans. These animals have a sensory capability far beyond our own. Some of them have senses that have no counterpart at all in human experience. These sensory systems may provide such animals with an awareness of the environment and a decision-making capability based on sensory qualities that we can no better appreciate than a congenitally blind human can appreciate a description of the range and subtlety of color shadings of a spectacular sunset. A dramatic example of such sensory capabilities are the vagal and facial lobes of carps and catfishes. These animals have an extraordinary gustatory system that receives information from taste receptors located not only in the mouth, but over most of the external body surface as well. The sensory innervation of these receptors terminates in a massive hypertrophy of the visceral afferent column of the medulla known as the vagal and facial lobes. The vagal lobes are laminated and have an overall organization that is in some ways reminiscent of the optic tectum. Other examples of highly developed hindbrain senses are the lateral-line lobes of electric fishes, which are capable of detecting the minute differences in the conductivity of the surrounding water that result from the presence of different types of living and nonliving objects, the infrared detectors served by the trigeminal nerve of some snakes that can detect a local temperature difference of 0.003 $^{\circ}C$ at a distance of 0.5 meter, as well as the superb auditory nuclei of echolocating birds and mammals. Do these highly developed and organized neuronal aggregations merely provide the animals with qualitatively and quantitatively superior sensory capabilities or do they participate in decision-making processes as well?

Cerebellum

The cerebellum is another structure that is usually eliminated readily from consideration because its overall organization is generally the same among the different vertebrate classes and its size seems to vary more with the animal's use of the vertical dimension of space rather than presumed intellectual abilities. Moreover, the effects of cerebellar damage in humans and animals rarely suggest cognitive or intellectual impairments. An exception to this generalization may be found in certain sharks and bony fishes which have achieved an extraordinary degree of cerebellar development. In these animals, the cerebellum is involved with their superior electroreceptive capabilities and may have taken on functions related to intelligent behavior that we would not suspect based on our knowledge of cerebellar function in mammals.

Mesencephalic Tectum

The tectum of the midbrain shows a considerable degree of variation among vertebrate species. Although its overall organization is roughly the same in all, its degree of development and differentiation differs markedly. A common characteristic of this structure appears to be that it processes a topographic representation of the external environment based on input from distance receptors (vision, audition, infrared, etc.). Thus, the tectum may be involved in the organism's awareness of the spatial properties of its external environment and their correlation with its own body image, which is also represented in the tectum.

Diencephalon

The diencephalon of anamniotes is generally less well differentiated than its amniote counterpart. In general the boundaries of the various nuclear groups are more difficult to visualize than in amniotes. Among the amniotes, the diencephalon of reptiles and birds is well differentiated into easily recognizable nuclear masses. A considerable number of these have been described as being comparable to specific diencephalic nuclei of mammals largely on the basis of their afferent and efferent relationships with other cell groups. Presumably, cognitive or intellectual functions that can be correlated with some property of diencephalic nuclei in mammals should have a rough counterpart in the corresponding nuclear group in birds and reptiles. An additional point to consider is that the hypothalamus of a number of anamniotes is relatively larger than that of amniotes and appears to contain specialized functions not found in amniotes. To what extent these functions may be more than merely vegetative and instead related to the animals' ability to behave adaptively in response to changes in environmental demands remains to be seen. However, the possibility should not be ruled out.

Telencephalon

The range of variation of the size and degree of differentiation of the telencephalon is considerable among the vertebrate taxa. These differences have been the basis of most of the early speculation about the evolution of the brain and its relationship to the evolution of intelligent behavior. Unfortunately, much of this speculation was based on inadequate data and a once-popular model of brain evolution that is not compatible with the data of contemporary comparative neuroanatomy (1, 9, 10, 15, 20, 24). This model postulates that only mammals have a neocortex; i.e., the cortex is "neo-" to mammals. Moreover, this model states that the telencephalon of nonmammals consists of a massive corpus striatum. When neocortex evolved in mammals, it took over the functions that were earlier performed by the striatum. In sharp contrast to this view, the present evidence suggests strongly that the same basic subdivisions of the telencephalon are present in all vertebrates and that cerebral evolution has been much more conservative than the early comparative neuroanatomists realized. Moreover, the proportion of the telencephalon that is comparable to the mammalian corpus striatum occupies roughly the same proportion of the telencephalon in all vertebrates. The remainder of the telencephalon appears to contain the same cell populations, with the same basic pattern of ascending and descending pathways as are present in mammals. The mammalian telencephalon is characterized by a very compact laminar organization of cells and fibers along the surface. The non-mammalian telencephalon is also organized into laminae, but these are not nearly as compact as the mammalian laminae and for many years their laminar organization was not recognized.

To be sure, substantial differences exist among vertebrates in the degree of development and differentiation of their telencephalon and the type of neuronal organizations that are present. The contemporary view of these differences is that they are in part due to phyletic differences and in part due to the degree to which the animal in question is adapted to make use of various telencephalic subdivisions in its interactions with its environment. However, these differences exist within an overall organizational framework that is common to all vertebrates. The differences manifest themselves in cell number, synaptic and dendritic organization, and the number and types of interconnections with other cell populations. Some of these interconnections are with remote cell groups and some are with cells that are quite proximal. The latter are known as local circuits. The greater the degree of lamination the more efficiently these local circuits are presumed to function. One might even speculate that the degree of development of these local circuits might be related to the cognitive life of the organism. If such be the case, then the degree of lamination might be an indicator of the extent to which a structure participates in behavioral processes that could be called cognitive. This hypothesis could be applied to regions of the brain other than the telencephalon such as the vagal lobes of fishes, the optic tectum, the olfactory bulbs of highly macrosmatic animals, etc. One of the many questions that this line of speculation raises is what is the minimal degree of lamination that would be necessary for a process to be cognitive?

Integrative Systems

The reticular formation and limbic system are major brain systems that link together cell populations from a number of brain regions. Although they have morphological and functional relationships with receptor and effector systems, they are quite separate from them. They have been implicated in sleep and wakefulness, arousal, attention, motivation, emotion, memory, information processing of various sorts, and other cognitive functions. Space limitations permit me to comment on only a few of the many components of these systems.

The reticular formation is present in all vertebrates, but is best developed in amniotes. Some anamniote counterparts of amniote reticular formation nuclei consist of clusters of only six or eight cells. The gigantic Mauthner cells, of which there are only one on each side of the midline, are a feature of the reticular formation of anamniotes. Among the amniotes, someone familiar with the reticular formation of mammals would feel quite at home studying the reticular formation of reptiles and birds.

Certain components of the limbic system are readily recognizable in all vertebrates, such as the septum and habenula. In many vertebrates, these structures are quite large and well developed. On the other hand, the hippocampus, while present (or at least a structure that has been called hippocampus is present) in all vertebrates, only in mammals does it achieve a massive size and high degree of differentiation and organization. Unfortunately, considerably less research has been done on the limbic system of non-mammals than has been done on sensory and motor systems and consequently considerably less information is available about the anatomical organization of its component cell groups and their interrelationships than in the sensory and motor systems.

CONCLUSIONS

If we are to find signs of intelligence in the animal kingdom and relate them to developments in neural structures, we must abandon the unilinear, hierarchical models that have dominated both searches. We must accept a more general definition of intelligence than one closely tied to human needs and values. We must accept the fact that divergence and nonlinearities characterize evolutionary history, and we must not expect to find smooth progressions from one major taxon to another. Finally, we must not allow ourselves to be biased by our knowledge of the mammalian central nervous system in our search for neural correlates of intelligence in other vertebrate classes. Without such changes in our thinking, we would appear to have little hope of progressing any further than we have in our attempt to understand the relationships between the human mind and the animal mind and their respective neural substrates.

Acknowledgement. Supported in part by grant number EY00735 from the National Eye Institute, U.S. Public Health Service. I am grateful to S.E.Brauth, C.J. Bartlett, C.B.G. Campbell, and I.T. Diamond for their valuable comments and suggestions.

REFERENCES

(1) Ariëns Kappers, C.U.; Huber, C.G.; and Crosby, E.C. 1960 (reprint of 1936 edition). The Comparative Anatomy of the Nervous System of Vertebrates, Including Man. New York: Hafner.

(2) Alcock, J. 1972. The evolution of the use of tools by feeding animals. Evolution 26: 464-473.

(3) Bitterman, M.E. 1975. The comparative analysis of learning. Science 188: 699-709.

(4) Boring, E.G. 1933. The Physical Dimensions of Consciousness. New York: Century.

(5) Buerger, A.A., and Fennessy, A. 19 . Learning of leg position in chronic spinal rats. Nature 225: 751-752.

(6) Campbell, C.B.G. 1976. What animals should we compare? In Evolution, Brain and Behavior: Persistent Problems, eds. R.B. Masterton, W. Hodos, and H. Jerison, pp. 107-114. Hillsdale, NJ: Lawrence Erlbaum Associates.

(7) Corning, W.J.; Dyal, J.A.; and Lahue, R. 1976. Intelligence: an invertebrate perspective. In Evolution, Brain and Behavior: Persistent Problems, eds. R.B. Masterton, W. Hodos, and H. Jerison, pp. 215-263. Hillsdale, NJ: Lawrence Erlbaum Associates.

(8) Dewsbury, D.A. 1978. Comparative Animal Behavior. New York: McGraw-Hill.

(9) Ebbesson, S.O.E. 1980. Comparative Neurology of the Telencephalon. New York: Plenum.

(10) Ebbesson, S.O.E., and Northcutt, R.G. 1976. Neurology of anamniotic vertebrates. In Evolution of Brain and Behavior in Vertebrates, eds. R.B. Masterton, M.E. Bitterman, C.B.G. Campbell, and N. Hotton, pp. 115-146. Hillsdale, NJ: Lawrence Erlbaum Associates.

(11) Gould, S.J. 1976. Grades and clades revisited. In Evolution, Brain and Behavior: Persistent Problems, eds. R.B. Masterton, W. Hodos, and H. Jerison, pp. 115-122. Hillsdale, NJ: Lawrence Erlbaum Associates.

(12) Hodos, W. 1970. Evolutionary interpretation of neural and behavioral studies of living vertebrates. In The Neuro-Sciences: Second Study Program, ed. F.O. Schmitt, pp. 26-39. New York: Rockefeller University Press.

(13) Hodos, W., and Campbell, C.B.G. 1969. Scala Naturae: Why there is no theory in comparative psychology. Psychol. Rev. 76: 337-350.

(14) Jerison, H.J. 1973. Evolution of the Brain and Intelligence. New York: Academic Press.

(15) Karten, H.J. 1969. The organization of the avian telencephalon and some speculation on the phylogeny of the amniote telencephalon. Ann. NY Acad. Sci. 167: 164-179.

(16) Llinás, R. 1969. Neurobiology of Cerebellar Evolution and Development. Chicago: A.M.A. Education and Research Foundation.

(17) Lloyd Morgan, C. 1894. Introduction to Comparative Psychology. New York: Scribner.

(18) Masterton, R.B.; Bitterman, M.E.; Campbell, C.B.G.; and Hotton, N. 1976. Evolution of Brain and Behavior in Vertebrates. Hillsdale, NJ: Lawrence Erlbaum Associates.

(19) Masterton, R.B.; Hodos, W.; and Jerison, H.J. 1976. Evolution, Brain and Behavior: Persistent Problems. Hillsdale, NJ: Lawrence Erlbaum Associates.

(20) Northcutt, R.G. 1981. Evolution of the telencephalon in nonmammals. Ann. Rev. Neurosci. 4: 301-350.

(21) Olson, E.C. 1976. Rates of evolution of the brain and behavior. In Evolution, Brain and Behavior: Persistent Problems, eds. R.B. Masterton, W. Hodos, and H. Jerison, pp. 47-78. Hillsdale, NJ: Lawrence Erlbaum Associates.

(22) Petras, J.M. 1976. Comparative anatomy of the tetrapod spinal cord: dorsal root connections. In Evolution of Brain and Behavior in Vertebrates, eds. R.B. Masterton, M.E. Bitterman, C.B.G. Campbell, and N. Hotton, pp. 345-381. Hillsdale, NJ: Lawrence Erlbaum Associates.

(23) Pearson, R., and Pearson, L. 1976. The Vertebrate Brain. New York: Academic Press.

(24) Prosser, C.L. 1973. Comparative Animal Physiology. Philadelphia: Saunders.

(25) Riopelle, A.J., and Hill, C.W. 1973. Complex processes. In Comparative Psychology: A Modern Survey, eds. D.A. Dewsbury and D.A. Rethlingshafer, pp. 510-546. New York: McGraw-Hill.

(26) Romanes, G.J. 1883. Animal Intelligence. New York: Appleton.

(27) Tuddenham, R.D. 1963. The nature and measurement of intelligence. In Psychology in the Making: Histories and Selected Research Problems, ed. L. Postman, pp. 469-525. New York: Knopf.

(28) Wallace, R.A. 1979. Animal Behavior: Its Development, Ecology and Evolution. Santa Monica, CA: Goodyear.

(29) Warren, J.M. 1976. Tool use in mammals. In Evolution of Brain and Behavior in Vertebrates, eds. R.B. Masterton, M.E. Bitterman, C.B.G. Campbell, and N. Hotton, pp. 407-424. Hillsdale, NJ: Lawrence Erlbaum Associates.

Animal Mind - Human Mind, ed. D.R. Griffin, pp. 57-74.
Dahlem Konferenzen 1982. Berlin, Heidelberg, New York: Springer-Verlag.

Mental Processes in the Nonverbal Hemisphere

J. Levy
Dept. of Behavioral Sciences, University of Chicago
Chicago, IL 60637, USA

Abstract. Studies of patients with unilateral cerebral lesions or with commissurotomy and of normal individuals show that the two sides of the brain are functionally asymmetric. Although the right hemisphere manifests profound deficiencies in language, it is superior to the left hemisphere in tasks requiring a good understanding of spatial relationships or memory for experiences resistant to verbal description. However, in recent years, some have suggested that only the left hemisphere is conscious, and others have proposed that the right hemisphere only surpasses the left for tasks involving spatial/manipulative skills of the hands. Evidence bearing on these issues is reviewed, and its relevance for understanding the animal mind is discussed.

INTRODUCTION

The role of the right hemisphere in behavioral and psychological function has been a matter of controversy since the classical 19th century discoveries of Dax, Broca, and Wernicke relating speech and other linguistic capacities to left-hemisphere specialization. The suggestion of Hughlings Jackson that the right hemisphere might have special abilities of its own in the realm of certain perceptual skills was a striking exception to the generally held view that the right side of the brain was an unconscious automaton, irrelevant for sensory integration, motor planning, and all higher processes.

It was conceived to be a mere relay station, sending information from the left side of space to its thinking partner on the other side and conveying motor commands, originating in the left hemisphere, to the muscles on the left side of the body.

During the last half-century, studies of unilaterally brain-damaged patients, split-brain patients, and normal individuals have provided evidence that both sides of the brain have specialized capacities complementary to those of the other hemisphere. Yet, in recent years, the question has been raised as to whether all the interest in laterality research is anything more than a self-reinforcing fad, whether the various claims made for right-hemisphere function may represent only popularized fantasies. Sir John Eccles (6), although admitting that the right hemisphere manifests better skills in some domains than the left, denies that it is conscious. Consciousness and thought, in this perspective, are synonymous with language. Michael Gazzaniga and his students and associates (13) reject the view that the right hemisphere has any cognitive superiorities and, instead, hold that the right hemisphere gains a certain "manipulo-spatial" advantage of the left hand by default as the left hemisphere becomes progressively committed to language with development. This interpretation denies any inherent, evolved superiorities of the right hemisphere even in the "manipulo-spatial" domain, and restricts the manifest right-hemisphere capacities to those involving manipulation of objects with the hand.

The emphasis of these interpretations is that language is crucial for thinking and consciousness and, if correct, would imply that non-human animals are mindless automata. Any search for the characteristics of the "animal mind" would be futile from the start since the search would be predicated on a contradiction. It therefore becomes important for those interested in the possibility of animal mentation and the nature of the mental properties of animals to determine whether, in fact, the nonverbal hemisphere of the human brain is capable of thought and reasoning and whether it is inherently superior to the verbal hemisphere in important and interesting domains of cognition. Should the evidence support this conclusion, we would have no a priori grounds for denying cognitive processes to animals. Furthermore, if special abilities of the right hemisphere are innate and are not merely gained by default with the maturation of language, we would be led to consider whether the functional asymmetries of the cerebral hemispheres have a more fundamental distinction than verbal versus nonverbal and are also found in the brains of other animals, and for the same adaptive reasons as in people.

Because the current literature on hemispheric lateralization is now so vast, it is not possible to give a detailed citation in a brief communication. Excellent summaries may be found in Dimond and Beaumont (5), Hêcaen (10), Schmitt and Worden (23), and Springer and Deutsch (25).

"MANIPULO-SPATIAL" ABILITIES OF THE RIGHT HEMISPHERE

Many cognitive neurologists and neuropsychologists have noted an association between right-hemisphere lesions and constructional apraxia. In this disorder, patients manifest serious deficits in perspective drawing, copying of simple pictures or designs, and in constructing patterns from colored blocks; such apraxic disabilities are significantly less common with left-hemisphere lesions. In patients with left-side damage, there is often an oversimplication and lack of detail in drawings, but basic form relationships are typically preserved. In contrast, patients with right-side damage seem to be unable to represent basic form, although many disconnected details and features may be depicted.

During the early months following surgery when manual control is strictly contralateral, split-brain patients also manifest the same hemispheric differentiation in constructional tasks.

Drawings made by the right hand (left hemisphere), in spite of the fact that patients are right-handed, are considerably poorer than those made with the left hand (right hemisphere) with respect to accuracy of copying and accuracy of relationships among parts. Similarly, the right hand has serious difficulty in constructing block designs, but the left hand performs adequately or even excels. Of possible relevance, normal right-handers have a left-hand superiority in difficult static finger-positioning tasks, as Kimura and Vanderwolf discovered (12).

In addition to its apparent praxic constructional skills, there is considerable evidence for a right-hemisphere superiority in spatial tasks involving manual palpation of either stimulus or choice objects. Witelson (27), Cioffi and Kandel (2), and others have observed a left-hand advantage for tactile-visual matching of nonsense shapes in normal children, and Rudel and her associates (22) discovered a left-hand superiority for Braille letter identification, congruent with the predominant use of the left hand for Braille reading in the blind.

Split-brain studies confirm and extend these findings from normal populations. Although each hemisphere of the commissurotomy patient eventually gains motoric control over the ipsilateral hand, presumeably via the extrapyramidal and uncrossed pyramidal tracts, stereognostic information presented to one hand remains strictly contralaterally projected. Thus, patients are unable to identify verbally objects placed in the left hand and cannot cross-match objects felt by the two hands; stimulus information extracted by feeling objects with the left or right hand is known only by the right or left hemisphere, respectively.

If patients are given small wooden blocks for tactile inspection, varying in shape and/or in the relationships among smooth and rough surfaces, and are required to select a matching choice among "opened-up" 2-dimensional drawings in free vision, performance is greatly superior when stimuli are presented to the left hand. Robert Nebes (18,19) found a clear left-hand superiority for tactile-visual, visual-tactile, and tactile tactile matching of an arc with a circle of the same diameter. No hand asymmetries emerged for either arc-arc or for circle-circle matching, indicating that the left-hand advantage for arc-circle matching could not be due merely to a manipulative or perceptual superiority of the left hand. In a later study, Nebes presented fragmented figures visually in an experimental condition and unified figures in a control condition, with the patients' task being to select by touch a shape representing the unification of the fragmented figure or identical to the unified figure. The left hand performed much more accurately than the right in the figural unification task and there was no hand asymmetry for the control task, although tactile choice objects were identical under the two conditions.

Laura Franco and Roger Sperry (9) investigated hemispheric differences in split-brain patients for identifying a geometric or topological invariant among a set of 5 objects displayed visually. Patients selected by touch, using either the left or right hand, an object matching the visual set with respect to shared Euclidian, affine, projective, or topological characteristics, with defining spatial constraints diminishing progressively through the 4 types of sets. The left hand surpassed the right hand on all sets and displayed a similar level of accuracy for the 4 kinds of tasks. The right hand performed best on Euclidian sets and progressively worse on affine, projective, and topological sets, with performance at chance level on the latter. The differences in patterns of performance of the two hands over the 4 types of sets suggests differences in the strategies of processing applied by the two hemispheres.

The various investigations are most easily interpreted as indicating a superiority of the right hemisphere for understanding and memory of spatial relationships and stimuli resistant to verbal description. A simple "manipulo-spatial" advantage seems inadequate to characterize the specializations of the right hemisphere. If the right hemisphere is at a cognitive advantage for certain types of functions, this should be manifest in tasks not involving manual manipulation.

SUPERIORITIES OF THE RIGHT HEMISPHERE NOT INVOLVING MANUAL SKILLS

In addition to the constructional apraxias, patients with right-hemisphere lesions display disorders in perceptual analysis and synthesis. McFie and colleagues (17) found that right- but not left-hemisphere damage leads to deficiencies in cube-counting performance. In this task, a stack of cubes is depicted in a drawing, with some cubes hidden, and it is the patients' task to determine the number of cubes in the stack. Successful performance depends on the ability to visualize how the cubes are arranged in 3-dimensional space. Hécaen and others have also shown that loss of the ability to recognize faces is almost never observed in patients with unilateral left-hemisphere lesions, but is not infrequently observed in patients with right-hemisphere or bilateral damage. Even when complete facial agnosia is not found, poor performance on face memory tests is common with right-sided damage.

Warrington and her colleagues (26) compared patients with left- and right-hemisphere lesions for the ability to recognize incompletely drawn letters or figures, judgements of dot location and line orientation, and comparative judgements of the size of a gap in simple figures, finding the right-lesion group to be significantly inferior on all measures, and particularly for patients with lesions of the parietal cortex. Spinnler and his associates (8,24), in several studies, found that right-hemisphere lesions, to a significantly greater extent than left-hemisphere lesions, result in poor perception and memory for subtle color variations among non-nameable colors.

The apparent perceptual superiority of the right hemisphere, as inferred from lesion studies, has been confirmed in split-brain patients. Nebes presented arrays of dots tachistoscopically to the left or right hemisphere, and patients were asked to indicate whether the dots were aligned in columns or in rows (defined by relative dot distances along horizontal versus vertical axes). Accuracy was significantly higher for arrays seen by the right hemisphere, and asymmetry of performance in favor of the right hemisphere increased as the relative column-versus-row dot distances decreased. These data suggest a more refined spatial metric for the right hemisphere and/or a better developed Gestalt organizational capacity.

Levy, Trevarthen, and Sperry (16) administered tests of competitive perception to commissurotomy patients, where two different stimuli were simultaneously tachistoscopically presented to the two hemispheres. In one condition, patients were asked to point to a matching stimulus in free vision and in the other condition, they were asked to name or describe what they saw. As expected, when a verbal response was required, the stimulus seen by the left hemisphere was described; the right hemisphere behaved as if it had seen nothing at all. When a matching response was required, although both hemispheres were capable of controlling the pointing response, matches were made to the right-hemisphere stimulus. The right hemisphere dominated matching for faces, nonsense shapes, pictures of common objects, and vertically oriented 3-element arrays of X's and squares. The matching performance of the right hemisphere was superior to the naming performance of the left hemisphere for faces and nonsense shapes. However, both hemispheres performed with essentially perfect accuracy on pictures of common objects, and the verbal performance of the left hemisphere surpassed the matching performance of the right hemisphere for the 3-element arrays.

Evidently, the instruction to perform an exact visual match alerted and activated the right hemisphere, both for tasks in which it surpassed the left hemisphere (faces and nonsense shapes) and for tasks in which the hemispheres were equal or the left hemisphere was superior (common objects and 3-element arrays). Subsequent competitive perception tests showed that the right hemisphere dominates behavior when patients are instructed to match for visuo-structural similarity of stimulus and choice and that the left hemisphere dominates matching behavior when words have to be matched to pictures, when pictures have to be matched for rhyming names, or when stimulus and choice are to be matched according to functional category (e.g., a hat with gloves).

The poorer performance of the right hemisphere for remembering 3-element arrays may result from the absence of "visual syntax" in these stimuli. The 3-element patterns differ from most visual stimuli in that arrangement of features (the X's and squares) is random across stimuli and subject to no constraints deriving either from overall form or from meaning.

Deductions drawn from investigations of neurological patients have been strongly confirmed in normative studies. In tachistoscopic research, a left-visual-field/right-hemisphere superiority has regularly been observed for matching or identifying faces or facial emotions, for discriminating line orientations having no verbal designations, for dot localization, dot enumeration, nonsense shapes that are rotated with respect to choice objects, stereoscopic depth discrimination, matching of non-nameable colors or shades of gray, and other tasks dependent on good spatial representation or imagistic memory. In dichotic listening tests, a left-ear (right hemisphere) advantage has been found for musical chords and nonverbalizable environmental sounds. Of relevance to the question of whether right-hemisphere skills derive from inherent organizational properties or are gained by default with development, infants have a right-ear advantage for discriminating consonant-vowel syllables and a left-ear advantage for discriminating musical timbre, as found by Entus (7) and by Best and Glanville (1).

Activational differences have also been observed between the hemispheres in normal people. The EEG alpha rhythm is selectively suppressed over the left hemisphere when subjects engage in verbal tasks and over the right hemisphere when they engage in spatial tasks. This same selective suppression is seen in infants with respect to a 4 Hertz rhythm that is the infant's homologue of the adult alpha: verbal stimulation results in asymmetric suppression over the left hemisphere and musical stimulation results in asymmetric suppression over the right hemisphere. Auditory evoked potential studies show larger evoked responses over the left hemisphere in response to verbal signals and the right hemisphere in response to nonverbal signals in adults, babies a few months of age, and neonates less than 24 hours of age. Recent investigations of regional cerebral blood flow using emission tomography reveal that there is an asymmetric increase of blood flow to the right hemisphere when subjects perform spatial tasks and asymmetric increase of blood flow to the left hemisphere when subjects perform verbal tasks.

DEFICIENCIES OF THE RIGHT HEMISPHERE

The deficiencies of the right hemisphere are as revealing as the superiorities in elucidating the nature of its specializations. Both receptive and expressive aphasias are found after left- but not right-hemisphere damage in the large majority of right-handers, although evidence from both split-brain patients and normal people shows that the right hemisphere can extract meaning from at least certain classes of spoken or written words. From analyses of reaction times for lexical decision tasks or for deciding whether a tachistoscopically presented word is an exemplar or not of a given superordinate category, James Day (3) was able to infer that the right hemisphere in normal people accurately classifies concrete nouns and adjectives presented in the left visual field, but not verbs or abstract nouns and adjectives.

Investigations of split-brain patients by Hillyard and Gazzaniga (11), Levy and Trevarthen (14,15), and Zaidel (29,30) show that although the right hemisphere can generally follow simple, redundant oral instructions where active-voice structure and a restricted concrete vocabulary is used and has a remarkably good comprehension vocabulary, it suffers profound disabilities in phonological and syntactical analysis. In contrast to its capacities for selecting written words to designate the name of objects it sees or as matching words it hears, the right hemisphere is completely unable to select written consonants as indicating the initial sound of spoken consonant vowel syllables, cannot match written homophones with different spelling patterns for identity of sound, and shows great defects in indicating whether or not a spoken word rhymes with the name of a picture it sees. It cannot distinguish singular from plural, past from present tense, or subject from object as given by syntax. Major disabilities are observed in decoding the meaning of spoken instructions on the Token Test either when syntax is complex or when a significant load is placed on verbal short-term memory. The "language" of the right hemisphere is radically different from that of the left, and indeed, it is not clear that, for the right hemisphere, spoken or written words differ in any qualitative way from other acoustic or visual stimuli for which associative meaning has been derived.

Beyond its language disorders, the right hemisphere is also deficient in making temporal-order judgements. Left-hemisphere damage leads to disruptions in the ability to judge the order of a pair of stimuli, whereas no effects are observed after right-hemisphere damage. Similarly, in normal people, the left hemisphere surpasses the right in judging the order of a pair of visual, auditory, or tactile stimuli, and the evidence suggests that it is the left hemisphere that performs temporal order analysis, even for stimuli presented in the left sensory field. The superiority of the left hemisphere in this domain is consistent with, and possibly critical for, its skills in analyzing phonology, phonemic order, and in decoding and organizing syntactically complex sentences. The right hand's superiority for writing, regulated tapping, or for any ordered series of movements may depend on a highly developed temporal-order sense. The ideational apraxias observed after left-hemisphere lesions, in which timed, ordered movements of the hand are disrupted, may reflect damage to underlying programs for analyzing and constructing time-ordered series.

The various superiorities of the right hemisphere appear in a framework of deficiencies in fundamental linguistic operations, sequential movements of the hand, and in temporal-order judgement, and it may be that right-hemisphere skills rely on programs of neural organization that are intrinsically incompetent for the cognitive specializations of the left hemisphere. Conversely, the neural organization underlying the left hemisphere's special abilities may be inherently incompetent for the specializations of the right side of the brain.

UNRESOLVED ISSUES AND QUESTIONS

Although the evidence, as currently available, gives strong support to the view that each side of the human brain has its own set of special processes that are complementary to those on the other side, that each has high-level cognitive skills in particular domains of thought, and that these differential abilities are programmed at birth and will emerge with normal maturation, there still remain a number of unanswered questions. It seems evident that language is not essential for thought, but why are various functions co-lateralized as they are? Why is it typically the case that when one hemisphere is specialized for speech, it is the other hemisphere that is specialized for face recognition? What is the nature of the underlying programs of organization that result in the particular forms of asymmetry observed?

One possibility is that the type of neural computer - its hardware and programming - needed for temporal-order judgements, analysis of feature characteristics of stimuli, and control of temporally ordered movements of the vocal apparatus and refined activities of the fingers and hands is ill-designed for the specializations of the right cerebral hemisphere, and conversely, that the neural organization underlying high-level skills in the spatial-perceptive domains is inadequate for left-hemisphere specializations. Lateral differentiation of function, regardless of its specific characteristics, would be expected to yield greater cognitive efficiency and power as compared to a perfectly symmetric brain simply by a deduplication of processes entailed by the reduction in redundancy, in the same way that the asymmetry of the viscera probably reflects a response to adaptive demands for the efficient use of internal bodily space.

There are two inferences that should follow from these interpretations. First, hemispheric lateralization should be observed in other animals, and, second, for those species whose adaptations are critically dependent both on temporal-order analysis and programming and on high-level spatial-perceptive abilities, opposite sides of the brain should be specialized for these two domains of function. In the last few years, investigations from a number of laboratories have, indeed, revealed anatomical and functional asymmetries in various groups of mammals, including rats, monkeys, and apes (4), and those observed in primates appear to be homologous with those in man. In the chimpanzee, as in people, the left Sylvian fissure is longer than the right (28), and in Japanese macaques, there is a right-ear/left-hemisphere advantage in discriminating conspecific calls (21).

There is evidence, also, for a parallel evolution of lateral asymmetry in certain songbirds. Nottebohm (20) has found that unilateral lesions of the hyperstriatum ventrale, pars caudale (HVc) in male canaries is more disruptive of song when produced in the left hemisphere than in the right. After right-hemisphere damage, birds continued to produce a significant fraction of their preoperative repertoire, and some birds sang as well as they had presurgically. Birds sang vigorously and their song was organized in clear-cut phrases. Nottebohm states that, "To the ear, these songs approached performance of intact birds."

In contrast, following left-hemisphere damage, there were serious disruptions in singing. Birds could produce none, or at most one, of the song syllables produced preoperatively; singing was very monotonous and was characterized by an unstable, simple succession of notes, rising and falling in pitch, with occasional incidences of two simultaneous and unrelated sounds.

Since both human speech and bird song entail a highly complex sequencing of sounds, the question arises as to whether the lateralization of these processes derives, for both groups, from similar adaptive demands, whether, in other words, these functions are asymmetrically dependent on one hemisphere so as to leave the other hemisphere free to develop an organizational program well-designed for spatial-perceptive processes. If so, one would expect to find that the right hemisphere of canaries surpasses the left in spatial localization and memory and in related cognitive operations. No data are yet available with respect to this issue, but if complementary specializations of the avian hemispheres were to be found, this would provide a strong suggestion that the forms of neural organization needed for temporal versus spatial capacities are sufficiently different to require lateralization for more optimal function.

Additionally, we still have little information regarding the underlying nature of lateralized functions in mammals and whether it is similar to that in people. Future research directed toward these questions can take us a long way toward understanding both the origins and psychological consequences of human cerebral asymmetry and the relationships between human and animal mental and neural function.

The cognitive complexity and competencies of the human right hemisphere, appearing in a framework of profound linguistic limitations, demonstrate that mentation is not dependent on the kinds of language skills of the human left hemisphere. Furthermore, it is difficult to assign any meaning whatsoever to a term like "consciousness" if we deny that the mental operations of the right side of the brain are based on conscious processes, and if, therefore, we are compelled to attribute consciousness to the nonverbal hemisphere, it seems that we are similarly compelled to admit the reality of the animal mind.

REFERENCES

(1) Best, C.T., and Glanville, B.B. 1978. Cerebral asymmetries in speech and timbre discrimination by 2-, 3-, and 4-month-old infants. Paper presented at the First International Conference on Infant Studies, Providence, RI.

(2) Cioffi, J., and Kandel, G. 1979. Laterality of stereognostic accuracy of children for words, shapes, and bigrams: a sex difference for bigrams. Science 204: 1432-1434.

(3) Day, J. 1977. Right-hemisphere language processing in normal right-handers. J. Exp. Psychol. Human Percep. Perform. 3: 518-528.

(4) Denenberg, V.H. 1981. Hemispheric laterality in animals and the effects of early experience. Behav. Brain Sci.: in press.

(5) Dimond, S.J., and Beaumont, J.G., eds. 1974. Hemisphere Function in the Human Brain. New York: John Wiley & Sons.

(6) Eccles, J.C. 1973. Brain, speech, and consciousness. Naturwiss. 60: 167-176.

(7) Entus, A.K. 1977. Hemispheric asymmetry in processing of dichotically presented speech and nonspeech stimuli by infants. In Language Development and Neurological Theory, eds. S.J. Segalowitz and F.A. Gruber. New York: Academic Press.

(8) De Renzi, E., and Spinnler, H. 1967. Impaired performance on color tasks in patients with hemispheric damage. Cortex 3: 194-217.

(9) Franco, L., and Sperry, R.W. 1977. Hemisphere lateralization for cognitive processing of geometry. Neuropsychologia 15: 107-114.

(10) Hécaen, H., ed. 1978. La Dominance Cérébrale: Une Anthologie. Paris: Mouton.

(11) Hillyard, S.A., and Gazzaniga, M.S. 1971. Language and the capacity of the right hemisphere. Neuropsychologia 9: 273-280.

(12) Kimura, D., and Vanderwolf, C.H. 1970. The relation between hand preference and the performance in individual finger movements by left and right hands. Brain 93: 769-774.

(13) LeDoux, J.E.; Wilson, D.H.; and Gazzaniga, M.S. 1977. Manipulo-spatial aspects of cerebral lateralization: clues to the origin of lateralization. Neuropsychologia 15: 743-749.

(14) Levy, J., and Trevarthen, C. 1976. Metacontrol of hemispheric function in human split-brain patients. J. Exp. Psychol. Human. Percep. Perform. 2: 299-312.

(15) Levy, J., and Trevarthen, C. 1977. Perceptual, semantic, and phonetic aspects of elementary language processes in split-brain patients. Brain 100: 105-118.

(16) Levy, J.; Trevarthen, C.; and Sperry, R.W. 1972. Perception of bilateral chimeric figures following hemispheric deconnection. Brain 95: 61-78.

(17) MeFie, J.; Piercy, M.F.; and Zangwill, O.L. 1950. Visual-spatial agnosia associated with lesions of the right cerebral hemisphere. Brain 73: 167-190.

(18) Nebes, R.D. 1971. Superiority of the minor hemisphere in commissurotomized man for the perception of part-whole relations. Cortex 7: 333-349.

(19) Nebes, R. 1972. Dominance of the minor hemisphere in commissurotomized man in a test of figural unification. Brain 95: 633-638.

(20) Nottebohm, F. 1977. Asymmetries in neural control of vocalization in the canary. In Lateralization in the Nervous System, eds. S. Harnad, R.W. Doty, L. Goldstein, J. Jaynes, and G. Krauthamer, pp. 23-44. New York: Academic Press.

(21) Petersen, M.R.; Beecher, M.D.; Zoloth, S.R.; Moody, D.B.; and Stebbins, W.C. 1978. Neural lateralization of species-specific vocalizations by Japanese Macaques (Macaca fuscata). Science 202: 324-327.

(22) Rudel, R.; Denckla, M.; and Spalten, E. 1974. The functional asymmetry of Braille letter learning in normal sighted children. Neurology 24: 733-738.

(23) Schmitt, F.O., and Worden, F.G., eds. 1974. The Neurosciences: Third Study Program. Cambridge, MA: The MIT Press.

(24) Scotti, G., and Spinnler, H. 1970. Colour imperception in unilateral hemisphere-damaged patients. J. Neurol., Neurosurg., Psychiat. 33: 22-28.

(25) Springer, S., and Deutsch, G. 1981. Left Brain, Right Brain. San Francisco: W.H. Freeman and Co.

(26) Warrington, E.K., and Rabin, P. 1970. Perceptual matching in patients with cerebral lesions. Neuropsychologia 8: 475-487.

(27) Witelson, S.F. 1974. Hemispheric specialization for linguistic and nonlinguistic tactual perception using a dichotomous stimulation technique. Cortex 10: 3-17.

(28) Yeni-Komshian, G.H., and Benson, D.A. 1976. Anatomical study of cerebral asymmetry in temporal lobe of humans, chimpanzees, and rhesus monkeys. Science 192: 387-389.

(29) Zaidel, E. 1976. Auditory vocabulary of the right hemisphere following brain bisection or hemidecortication. Cortex 12: 191-211.

(30) Zaidel, E. 1976. Language, dichotic listening, and the disconnected hemispheres. In Conference on Human Brain Function, eds. D.O. Walter, L. Rogers, and J.M. Finzi-Fried. Los Angeles: Brain Information Service/BRI Publications Office.

Animal Mind - Human Mind, ed. D.R. Griffin, pp. 75-94.
Dahlem Konferenzen 1982. Berlin, Heidelberg, New York: Springer-Verlag.

Risk-benefit Assessment in Animals

R. H. Drent
Zoölogische Laboratorium, Rijksuniversiteit Groningen
9751 NN Haren (Gr.), Netherlands

Abstract. How do animals themselves assess risks and benefits attached to the alternatives open to them? The foraging context has been chosen for detailed consideration on account of higher fitness presumed to accrue to efficient foragers, and discussion is limited to birds on account of the limited experience of the author. First, evidence is presented concerning the minimum type of information to which birds appear to be responsive in adjusting their foraging effort. Most of the decisions met with in the field are seen as outcomes of learning (memory). It is further argued that evaluation of when to abandon certain prey types, or in more general terms the decision how to allocate time (and effort) to alternative food supplies, may depend on monitoring the rate of energy expenditure (or state of energy reserves, i.e., net energy expenditure) even on a rather short time scale. In this sense, what may be termed the "first order" decisions in the foraging hierarchy have a physiological substrate. Similar considerations may guide the parents in reaching so-called "strategic" decisions (when to breed, how many eggs to lay, when to abandon a breeding attempt). "Mental competence" of the animal remains an elusive property.

INTRODUCTION

Although many will agree with Maynard Smith's contention that "Ecology is still a branch of science in which it is usually better to rely on the judgement of an experienced practitioner than on the predictions of a theorist" (20), there is no shortage of predictions in the current literature. In some areas at least the process of testing predictions has even penetrated to the ranks of field workers, traditionally prone to accept a passive observer's role. The application and extension of foraging theory as so masterfully propounded by MacArthur and his school (summary (17)) is a case in point and holds the promise that ecology will eventually come of age after all. I believe part of the answer to this success story (15) lies in the way in which MacArthur dissected the many facets of prey and environment that the successful forager needed to contend with by reducing them to a short but logical sequence of decisions. The weighing of alternatives is described as an assessment of the risks and benefits attached to them, and as natural selection is a hard school, it has become general to presuppose that animals operate optimally under the prevailing circumstances. Optimality criteria can be devised given the insight of "judgement of the experienced practitioner" and some at least have proven remarkably resilient to more than a decade of testing (Royama's (25) "profitability" concept, i.e., the maximization of the net energy intake, is a case in point). Agreement between empirical data and the predictions derived from "optimal foraging theory" is thus the first step in confirming the correct identification of the cost functions involved (in general, *intake maximizing* models prevail, see (15)). This procedure opens at least two approaches to an insight into the evaluation procedure followed by the animal itself (where cognitive processes may be involved). One line of attack is to develop the theory of the economics of decision making, deploying mathematical expertise in handling observations of animals in the confines of artificial arrangements simplifying the context of decisions, or even in the intricacy of the field (19). This approach suffers from the serious drawback, as admitted by one of its proponents, "of the sheer difficulty of understanding what has been done and the assumptions on which it is based" (18), so the circle following this research strategy is necessarily small as yet. For the more naive wishing to grapple with the decision-making process itself, there is ample scope in the search for the proximal mechanisms underlying choice. Even in such a well-worked field as foraging, few mechanisms touching on presumed mental faculties have been hypothesized. The most general is the "search image hypothesis" propounded almost parenthetically some thirty years ago (33) and still insufficiently understood (9) to provide more than a loose framework characterizing this particular form of learning as an adaptive answer to limitations of mental competence of the predator.

What goes on in the animal's mind as it crosses the successive hurdles in the hierarchy of decisions regulating parental feeding in birds? My approach will be to ask the following question: For which types of knowledge do we have compelling field evidence that birds weigh them in reaching their final decisions? This is an approach to the task of delimiting, however sketchily, the criteria the bird might use in organizing its behavior under natural conditions.

TIME

Close observation of foraging individuals has disclosed an ability of predators to return to previously visited sites in a systematic fashion. And in some cases it has been shown that the mean return interval maximizes the expectation of yield. Examples are nectarivorous birds that make circuits between flowers at a rate allowing replenishment of nectar (7) and the systematic return of wagtails to scavenge the debris at the edge of a stream (3). These are short-term rhythms (measured in fractions of an hour) bringing the predator back at about the time that yields maximum availability of prey. The ability to link such rhythms to periodic events in the environment is a more complex ability. The exploitation of tidal flats by shorebirds has been analyzed by experiments with captive oystercatchers utilizing a food source subject to a tidal regime. (That is, food was made available one hour later on each successive day, corresponding to the local tides.) These experiments have demonstrated an inherent timing sense in the predator, revealed by anticipatory behavior just before the expected time of low water (6). Field observations confirmed that these birds behaved as though they had full knowledge, i.e., predictive information, about the local tide tables. For they habitually arrived "early" on days when winds delayed the time of low water. They also arrived late on days when, under the influence of the wind, the flats were exposed sooner than might have been predicted.

A circadian rhythm of emergence in many insects may have evolved to reduce the probability of capture of individuals which emerge at the same time as their conspecifics. This in turn is mirrored in a circadian rhythm of prey capture ((22), as illustrated in Fig. 1). In the case of the Starling, laboratory work has demonstrated the ability of the bird to respond at the appropriate times of day to a variety of stimuli, each geared to a specific time (see Fig. 2). Reliance on a circadian template provides the simplest explanation for the observed periodicity in hunting activity of Kestrels when preying upon voles. These

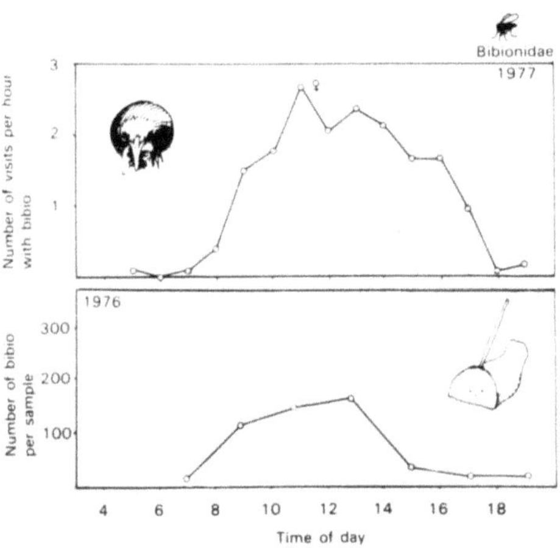

FIG. 1 - Exploitation of the emergence of Bibio flies by a parent Starling: the circadian rhythm of emergence (below, measured by sweep-netting) and the rate at which Biobinid flies were brought to the nestlings. The nest was monitored during a ten-day period by a film camera, enabling identification of prey (32).

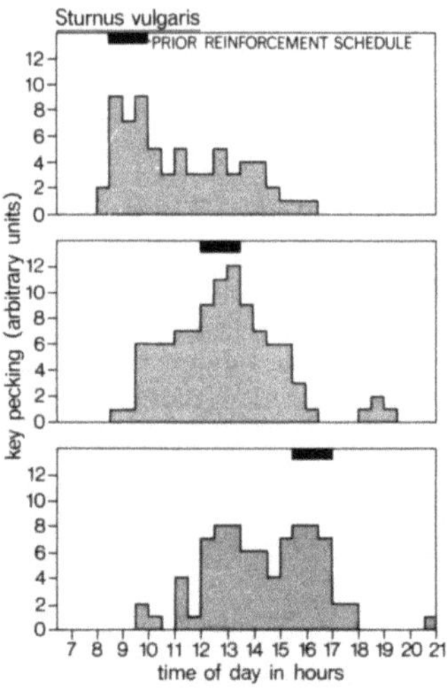

FIG. 2 - Allocation of pecks of three different keys by a captive Starling on test days (no reward) following training with time-restricted reward schedules as indicated by bars at top of each graph (5).

small rodents have a complex short-term rhythm of their own tied with their digestive machinery which will not concern us here, but the result is a synchronized peak in above-ground activity at intervals of 2-3 hours yielding peaks in vulnerability predictable on a circadian basis that the Kestrels learn to exploit (see (28)). As a result, the hunting activity of the Kestrel is closely synchronized with the vole activity cycle (Fig. 3). The circadian memory of the predator can be visualized as a tape loop (Enright in (5)) and the predator adjusts his allocation of hunting time to peak at times yielding high returns on previous days. Vital to the hypothesis is the finding that none of the individual Kestrels studied learned to exploit each of the vole peaks, but in fact "missed" some of the periodicity in their environment on account of presumed vagaries in their hunting on previous days, hence ruling out the development of a "wave-riding" tactic dependent on sensitivity of the interval between successive peaks.

Finally, several herbivores have been found to exhibit a periodicity in returning to food sources. Brent geese return every 4-5 days when exploiting plants in an actively growing phase, the timing depending on the rate of regrowth in relation to the cropping exerted by the geese (23). These observations with marked individuals support the speculations advanced earlier by Cody (3) concerning return times to food sources that replenish themselves. The interesting feature in the case of the geese is that the pattern of repeated cropping maintains the sward in a growth phase providing high protein content coupled with maximal digestibility. This enhancement of food quality by actions of the flock indicate how critical it is for the group members to revisit the sites at the appropriate interval. Measurements of intake in relation to standing stocks force one to the interpretation that the food supply is effectively depleted at each periodic visit.

In short, predators acquire information about periodicities in prey vulnerability, or in prey biomass and quality, and they can learn to react to these periodicities in an anticipatory

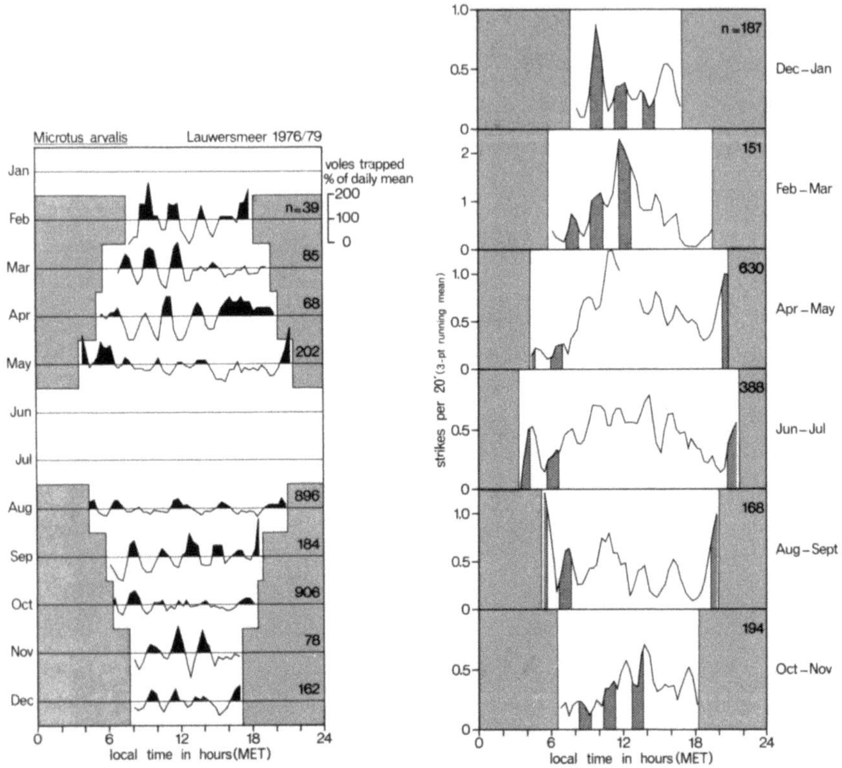

FIG. 3 - Hunting rhythmicity of the Kestrel in relation to above-ground activity rhythms in its main prey, the common vole. A calender of daily variation of voles caught in live-traps is given at left (number of voles trapped per 20 minutes expressed in relation to the mean for the daylight period and shown as 3-point running means; each monthly graph based on 2-9 trapping days, n = total voles caught). On the right, daily curves of strike frequency of Kestrels in the same area are shown, derived from continuous watches (n = hours of observation). For comparison, shaded bars indicate the vole activity peaks generalized from the trapping data on left (Assembled from (28)).

fashion. This is the case not only when the periodicities are anchored in a circadian template but equally when the periodicity shows some lag, as in tidal rhythms. Shorter-term periodicities (returning within the hour) must be considered together with a topographic component of food availability (see below). Longer-term rhythms (the grazing peaks, for instance) have not yet been analyzed experimentally. The circadian sensitivity of the Kestrels is reminiscent of the peaks in learning performance of rats at 28, 48, and 72 hr following the original training time as reported by Holloway and Wansley (13). Accordingly, biological clocks in higher animals find their significance in the repetition of behavior patterns from day to day (5), and it is apposite to close with a striking example of such repeatability (see Fig. 4).

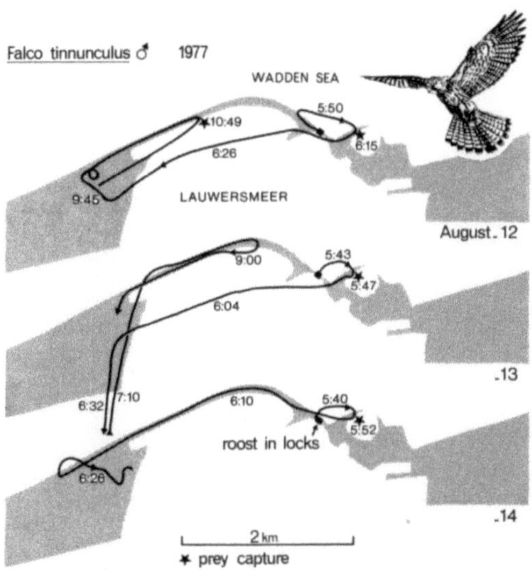

FIG. 4 - Daily routines as observed in a male Kestrel followed with the aid of telemetry on five successive dates (from (28)). Note the correspondence in time and site of vole captures (*) on three consecutive days.

LOCALITY

The separation of "when" from "where" is of course an artificial one, since every predator-prey interaction has both a time and an address. The observation that many predators revisit previous hunting areas periodically has generated field experiments in which a specific sector within the home-range of the individual concerned is baited (preferably an area previously characterized by a low yield). Starlings (30) and Kestrels (28) respond rapidly to a newly available food source providing high rates of return and reallocate time accordingly (see Fig. 5). An interesting feature of these experiments is that responsiveness to the baited area is specific to the time of day during which the original rewards were obtained, even after the supplementary food has been withdrawn. Clearly place-memory is an important attribute of the successful predator, especially when exacting demands are made on animals faced with diminishing food sources. With rare exceptions, an animal's feeding activities exert a measurable depleting effect on the subsequent rate of prey extraction at that site, most importantly because only a small fraction of the prey present are vulnerable to capture at any given moment. This avoidance of areas that are temporarily depleted, coupled with the possibilities of exploiting contagious distributions of catchable prey

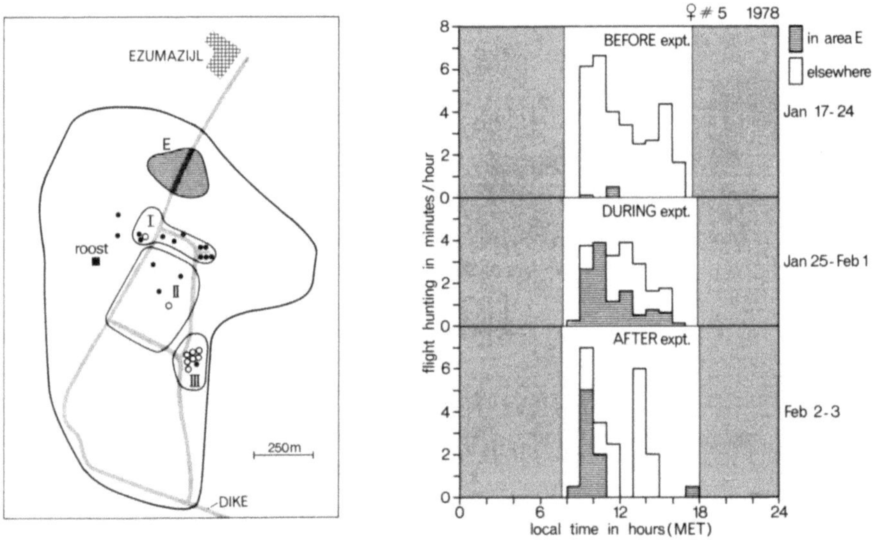

FIG. 5 - Manipulating the allocation of hunting time by a female Kestrel (from (28)). The map at left shows the home range and prey captures in January 1978 before the experiment (solid dots - prey taken in flight, open dots - from perches), and the experimental area (E) where prey were provided in the 8-day period January 25 - February 1 (the great majority between 09.00 and 10.00 hrs). Shown on the right is the dramatic effect of this supplementary feeding on allocation of hunting time, including a pronounced peak in visitation corresponding with the main feeding time after the supplementary feeding ceased (bottom panel).

items, require a precise topographic memory for the predator to be efficient. In fact, painstaking measurement of the landing sites of predators in relation to previous hunting experience points unequivocally to a highly developed use of topographic cues (31). When female geese feed on buds and other items that do not regenerate quickly (i.e., depleting a non-replenishing food source), and when particular individuals use the same sector of tundra for each brief foraging foray during the month-long incubation stage, they avoid crossing their tracks at least within the resolution of measurement. The retention of an "intake map" takes on a special significance when considered with regard to the choice between competing prey types.

Before going on to this complication, we can ask, how sensitive is the allocation of hunting time to the subunits utilized by the predator, and to variations in intake rate experienced? Experiments in which the rate of reward can be manipulated revealed that the probability of a predator making even small-scale shifts in areas visited is sensitive to single prey captures, or at most to one hunting session, even in predators that collect a huge number of prey over the course of the day (31). To the intricacies of the time-program of rewards familiar through the work of the experimental psychologists a topographic component is thus added. Not only must the predator have a "sliding memory window" of the patches recently visited (to calibrate the mean yield, see (15)), but to do justice to topographic complexity we must view the memory window at least as the many glittering panes of a palace.

PREY CHOICE

To cope with the intricacies of time and place dealt with so far, animals need only be equipped with memory (i.e., learning ability). But the choice criterion for discriminating between alternative prey types is more complex than intake maximization (or even energy minimization) alone. This generalization leans heavily on field experiments in the Starling parent when provisioning the young ((31), see Fig. 6). In the local situation

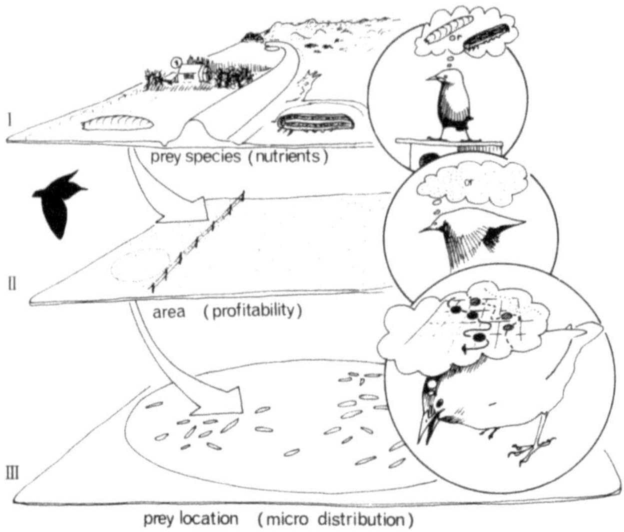

FIG. 6 - The hierarchy of foraging decision in parental Starlings as visualized by Tinbergen (31). The bird posed on the nestbox to depart on a collecting trip (top) must first decide which prey to hunt, since the alternative prey occur in different habitats. This choice is probably governed by nutrients, as mediated by the begging behavior of the young. In this example the bird chose the prey type on the left (leatherjacket). The second level of decision is where to land, and in this the bird relies on long-term information about expectation of yield in parts of the hunting area previously visited. The third level of decision (bottom) involves the exact landing site and subsequent search path, developed in relation to a fine-grained topographic memory of previous captures (tracks of successive visits are represented by decaying trajectories in a coordinate system).

where the study was carried out, two prey types are of paramount importance. One of them, the leatherjacket (Tipula), the Starlings find quickly and close to the colony, hence transport costs are low. The other, a caterpillar (Cerapteryx), not only demands a heavy investment in searching time, but also involves a much greater flight distance. When considered in terms of energy units delivered to the young for a given amount of parental time or energy devoted to collection and transportation of prey, bringing the caterpillars does not make sense at all. Nevertheless, choice experiments clearly prove that the parent Starling shows a strong preference for gathering caterpillars and persists in bringing a high proportion of this prey type unless brood demand outstrips collecting capacity (as can be demonstrated by manipulating the hunger state of the brood). Obviously the simplified theory of intake maximization is insufficient here, and we are forced to include "nutrient content" or some other description of the quality of the prey to interpret the parental decision. How does the parent know how to allocate its time between these two competing prey? Clearly the parent senses the hunger of the brood when it visits the nest to feed the nestlings, and the main problem to resolve is why, as brood demands increase, the balance of prey brought shifts in favor of the easy-to-get (but "low quality") leatherjackets. This adjustment can bring only temporary relief, since nestlings fed predominanty with leatherjackets suffer impaired survival. Where can the parent expect guidance in this dilemma? How far the parent should exert itself on behalf of the brood is of course a more general problem than the choice of prey type alone. On the simplest (proximal) hypothesis, there is a ceiling for the work capacity of the parent on a sustained basis, and thus (11,24), when confronted with enlarged broods, parents can be made to work up to that level, but not beyond (because their own survival is then impaired to such an extent that it is not compensated for by the possible survival of the current brood when weighed against the probability of raising another brood next season). Actual data on energy expenditure at the "Sustained Working Capacity" are scarce, but they have the virtue of clustering at a level corresponding to heavy labor by human standards (see Fig. 7). Does the introduction of this energetic constraint help us when we return to the problem of the parent Starling trying to titrate the amount of caterpillars in its collecting efforts?

Essential to testing the reality of ceiling values for sustained working capacity in the context of animal decision is the ability to monitor energy expenditure of individuals in the field. By incorporating the doubly-labeled water technique (2,21), we now have some information on the working levels of parent Starlings facing the prey selection dilemma outlined above. The simplest interpretation of the data is that the parent is always

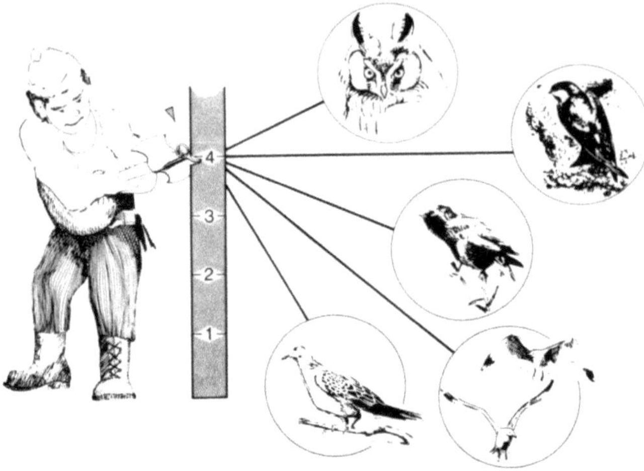

FIG. 7 - Maximum sustained working level of parent birds tending nestlings, expressed as metabolizable energy per day (DME) in multiples of basal metabolic rate (BMR). The points refer to Asio otus, Delichon urbica, Sturnus vulgaris, Larus glaucescens, and Streptopelia risoria, and cluster at approximately 4 BMR, the working level of heavy labor by human standards (11).

striving to deliver a maximum of the "high quality" (but low delivery rate) prey (the caterpillars) but delimits this maximum by remaining within a fairly narrow range of total energy expended per day. In other words, the contingent of "low quality" (but rapidly obtainable) prey is adjusted to provide "bulk food" for the nestlings, but the extent of this "topping up" of the nestling diet, necessary on account of the time cost involved in bringing the caterpillars, is limited by the capacity of the parent for sustained work. In working out the energy budgets it became clear that flight time was the major cost factor from the parent's point of view and an ability to monitor mean work levels when engaged in "runs" of collecting prey of the same type in the same general area would provide a rule of thumb adequate to allow the parent to adjust its effort among the alternatives. An evaluation procedure utilizing parental effort thus provides the optimal diet, in the sense of ensuring the diet richest in caterpillars within the physical reach of the parents under the prevailing circumstances. Assessment of the cost/benefit function for the parent in this context thus entails monitoring the state of the bird's own body reserves, and perhaps being subjectively responsive to a feeling of physical fatigue. The energy balance of the individual as a decision substrate has the attraction of wide applicability, as has been demonstrated on the basis of preliminary estimates of energetic assessment of the costs and benefits of the food function of territory (7). Recent modeling of parental decisions in voles (29) is also based on considering energy balance as the best measure of current fitness.

A further decision faced by the foraging parent is how many prey to bring back to the nest at a time (the "load size"), and this problem has been approached by considering the time savings involved in economizing on round trips in relation to the expected decline in collecting rate if the parent continues to hunt with prey already in its bill (22). There is a small discrepancy between the predicted and observed load sizes (31) in the direction to be expected if the parent is not assessing delivery rate alone (i.e., time basis) but includes effort expended (which tends to increase the "loading" the bird would apply to flight time in reaching its decision). To close this excursion into the foraging world of the parent bird, Tinbergen's concept of decision sequence is given in Fig. 8.

DISCUSSION: SHORTCOMINGS OF THE APPROACH

A problem faced when posing a decision problem in terms of a risk/benefit assessment is the time scale over which the risks or benefits are to be considered. In the foraging context, the optimal allocation of hunting time over the various segments of the home range can only be decided if we know the delivery rate that the parent must meet and the importance of acquiring information as a safeguard against failure (exhaustion) of the food supplies currently under exploitation. In words, we say the animal must "sample" the environment to ensure adequate future performance. That birds tend to "explore" more when not

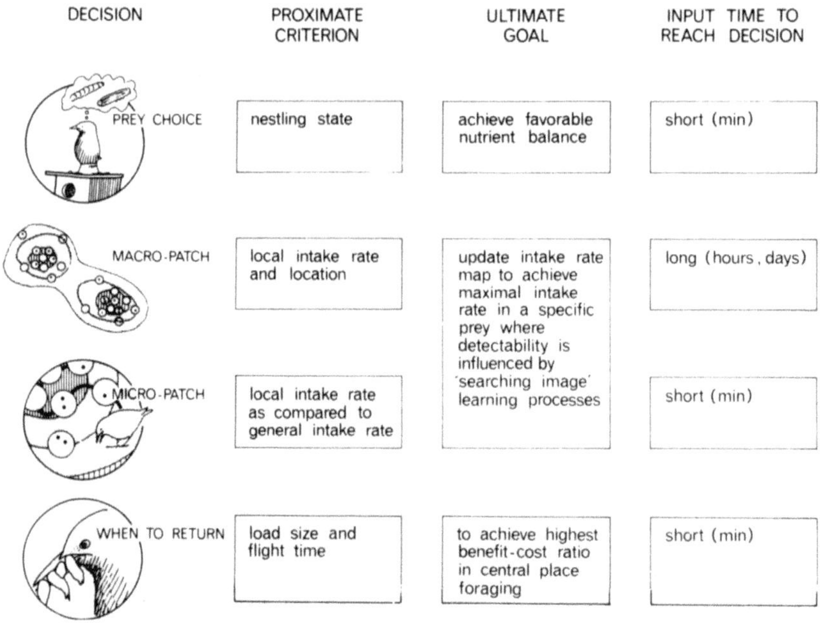

FIG. 8 - Tinbergen's (31) concept of the decision chain facing the parent Starling when provisioning young.

under pressure to maintain high delivery rates is one interpretation for the observed difference in foraging mode between parent and non-breeding individuals feeding in the same general area. Or, as Orians (22) puts it, the relatively relaxed non-breeders "were in a position to use their foraging times to learn things that might be useful in the future." Records of landings of parent Starlings when under differing pressures to collect food (i.e., differing levels of satiation of the young) can also be interpreted to support this "exploring for later" idea, as the birds under reduced pressure tended to avoid the precise vicinity of high-yield areas they had consistently returned to earlier (31). There is a great need for further field experimentation to bring these hazy outlines into focus. For laboratory work along these lines, however elegant it may be, need not necessarily lead to correct interpretation of decision making in nature.

A second unresolved problem in the risk/benefit approach is the calibration of "currency equivalence" between different functional contexts: we have carried on the discussion so far as if foraging alone dictates the animal's activities; but of course other functions such as predator avoidance will influence the final compromise reached. The theoreticians (19) and experimenters (14,16) are actively grappling with this issue, but considerable ingenuity will be required before observations in the field can become sufficiently comprehensive to provide an empirical approach to these complexities. Consider the repercussions of social status for foraging: P. Drent (personal communication) has found that not only is the relative ranking between individual Great Tits dependent on where the competition for food occurs, but when the birds meet in what might be termed a neutral ground, the route that the contestants have followed in getting there is of critical importance. An individual passing the scenes of previous defeats will assume a subordinate role, but if it has followed a route free from such associations, a new contest will be called for. In view of these subtleties, we will need to achieve a finer resolution in field observations of foraging before entering the arena of risk/benefit analysis interpreted in terms of fitness.

As far as I am aware, the only phenomenon in the observational data that has led to speculation on what is going on in the animal's mind concerns the increase in risk that a particular prey species will be captured by an individual predator, i.e., the observation of the sudden entry into the diet of a cryptic prey species (or morph of a species), unattended by changes in its density. L. Tinbergen (33) advanced the idea that birds may learn to see food that earlier was overlooked and termed this learning process the adoption of "specific searching images." Subsequent experiments both in birds (4) and fish (1) supported this idea of an enhanced discovery rate mediated by experience. De Ruiter, watching Jays kept in an enclosure stocked with various highly cryptic prey, found that a single episode of discovering that a certain twig-mimic (Hyloicus pinastri) was in fact palatable food, was sufficient to bring the item into the individual's diet in a steeply rising fashion (26), and this increased "awareness" of the prey was accompanied by the investigation and pecking of inedible objects of similar appearance, ignored before the experience with Hyloicus (27).

Dawkins (9) showed that the individual's ability to exploit cryptic prey declined if a period of experience with conspicuous prey were intercalated. Not only must the individual "get its eye in" when learning to see, but it must have repeated experience if it is to profit from its enhanced ability to discover the cryptic prey. (Any bird-watcher can attest to the importance of continuous exposure to "difficult" species to maintain quick recognition based on subtle characters.) There is good reason (9) to postulate central changes, and not the lack of input, as being responsible for the development of "specific searching images," and as naturalists we are forced to conclude that maintaining a heightened awareness of prey that are difficult to detect must have some cost, and hence is allowed to lapse into "forgetfulness" if not rewarded. That the discrimination of cryptic prey indeed entails a time cost has been recently demonstrated by Erichsen et al. (12) using an ingeneous conveyer belt technique of prey presentation, although even under these conditions it was not possible to measure recognition time in an absolute sense. We can thus sympathize with Orians who felt himself forced to adhere to MacArthur's simplification and "assume that the bird recognizes prey instantaneously and makes a choice between pursuing prey or ignoring it and continuing to hunt" (22). Nevertheless, this cannot be true, and until we have managed to collect the requisite measurements, a crucial part of the risk/benefit approach to interpreting an animal's diet will not be open to empirical test.

Acknowledgements. I thank my colleagues J.M. Tinbergen and S. Daan for their contributions towards this article. G. Thomas has been a helpful critic, and I am indebted to D.R. Griffin for smoothing out the final version.

REFERENCES

(1) Beukema, J.J. 1968. Predation by the Three-spined Stickleback (Gasterosteus aculeatus): the influence of hunger and experience. Behav. 31: 1-126.

(2) Bryant, D.M., and Westerterp, K.R. 1980. The energy budget of the House Martin (Delichon urbica). Ardea 68: 91-102.

(3) Cody, M.L. 1971. Finch flocks in the Mohave desert. Theor. Pop. Biol. 2: 142-148.

(4) Croze, H. 1970. Searching image in Carrion Crows. Z. Tierpsychol. Beiheft 5: 1-86.

(5) Daan, S. 1981. Adaptive daily strategies in behaviour. In Handbook of Behavioural Neurobiology: Biological Rhythms, ed. J. Ashoff, vol. 5, pp. 275-298. Heidelberg: Springer.

(6) Daan, S., and Koene, P. 1981. On the timing of foraging flights by Oystercatchers, Haematopus ostralegus, on tidal mudflats. Neth. J. Sea Res., in press.

(7) Davies, N.B. 1980. The economics of territorial behaviour in birds. Ardea 68: 63-74.

(8) Davies, N.B., and Houston, A.I. 1981. Owners and satellites: the economics of territory defence in the Pied Wagtail, Motacilla alba. J. Anim. Ecol. 50: 157-180.

(9) Dawkins, M. 1971. Perceptual changes in chicks: another look at the 'search image' concept. Anim. Behav. 19: 566-574.

(10) Drent, P.J. 1981. Dominance organisation in flocks of the Great Tit. Behav., in press.

(11) Drent, R.H., and Daan, S. 1980. The prudent parent: energetic adjustment in avian breeding. Ardea 68: 225-252.

(12) Erichsen, J.T.; Krebs, J.R.; and Houston, A.J. 1980. Optimal foraging and cryptic prey. J. Anim. Ecol. 49: 271-276.

(13) Holloway, F.A., and Wansley, R.A. 1973. Multiple retention deficits at periodic intervals after active and passive avoidance learning. Behav. Biol. 9: 1-14.

(14) Kacelnik, A. 1979. Studies on foraging behaviour and time budgeting in Great Tits (Parus major). Ph.D. thesis, Oxford University.

(15) Krebs, J.R. 1978. Optimal foraging: decision rules for predators. In Behavioural Ecology, An Evolutionary Approach, eds. J.R. Krebs and N.B. Davies, pp. 23-63. Oxford: Blackwell.

(16) Krebs, J.R. 1980. Optimal foraging, predation risk and territory defence. Ardea 68: 83-90.

(17) MacArthur, R.H. 1972. Geographical Ecology. New York: Harper & Row.

(18) McCleery, R.H. 1978. Optimal behaviour sequences and decision making. In Behavioural Ecology: An Evolutionary Approach, eds. J.R. Krebs and N.B. Davies, pp. 377-410. Oxford: Blackwell.

(19) McFarland, D.J. 1977. Decision making in animals. Nature 269: 15-21.

(20) Maynard Smith, J. 1974. Models in Ecology, pp. 1-146. Cambridge: Cambridge University Press.

(21) Nagy, K.A. 1980. CO_2-production in animals: analysis of potential errors in the doubly labeled water method. Am. J. Physiol. 238: R466-R473.

(22) Orians, G. 1980. Some adaptations of marsh-nesting blackbirds. Monogr. Pop. Biol. 14: 1-295. Princeton University Press.

(23) Prins, H.H.T.; Ydenberg, R.C.; and Drent, R.H. 1980. The interaction of brent geese Branta bernicla and sea plantain Plantago maritima during spring staging: field observations and experiments. Acta Bot. Neerl. 29: 585-596.

(24) Royama, T. 1966. Factors governing feeding rate, food requirement and brood size of nestling Great Tits (Parus major L.). Ibis 108: 313-347.

(25) Royama, T. 1970. Factors governing the hunting behaviour and the selection of food by the Great Tit (Parus major L.). J. Anim. Ecol. 30: 619-668.

(26) Ruiter, L. de. 1952. Some experiments on the camouflage of stick caterpillars. Behav. 4: 222-232.

(27) Ruiter, L. de. 1953. Camouflage in het dennenbos. I. De Dennepijlstaart. Levende Nat. 56: 41-48.

(28) Rijnsdorp, A.; Dijkstra, C.; and Daan, S. 1981. Hunting in the Kestrel, Falco tinnunculus, and the adaptive significance of daily habits. Oecologia, in press.

(29) Stenseth, N.C., and Framstad, E. 1980. Reproductive effort and optimal reproductive rates in small rodents. Oikos 34: 23-34.

(30) Tinbergen, J.M. 1976. How Starlings (Sturnus vulgaris L.) apportion their foraging time in a virtual single prey situation on a meadow. Ardea 64: 155-170.

(31) Tinbergen, J.M. 1981. Foraging decisions in Starlings Sturnus vulgaris L. Ardea 69: 1-67.

(32) Tinbergen, J.M., and Drent, R.H. 1980. The Starling as a successful forager. In Bird Problems in Agriculture, eds. E.N. Wright, I.R. Inglis, and C.J. Feare, pp. 83-97. London: British Crop Protection Council.

(33) Tinbergen, L. 1960. The natural control of insects in pine woods. I. Factors influencing the intensity of predation in song birds. Arch. Néerl. Zool. 13: 265-343.

Animal Mind - Human Mind, ed. D.R. Griffin, pp. 95-112.
Dahlem Konferenzen 1982. Berlin, Heidelberg, New York: Springer-Verlag.

The Ecological Conditions of Thinking*

D. Dörner
Lehrstuhl Psychologie II, Universität Bamberg
8600 Bamberg, F. R. Germany

Abstract. In this paper a system of hypotheses is presented about the integration of thinking processes into a general organizational schema of human behavior. The general organizational schema of behavior consists of three basic behavioral components and their interactions, namely, orientation behavior, monitoring behavior, and effect-directed behavior. The regularities of these forms of behavior are discussed. Thinking is the most developed form of effect-directed behavior. The root and core of human thinking is an internal simulation of external behavior. The conditions for the autonomy of thinking as internalized behavior are discussed. The standard of thinking is strongly dependent on the extent of control which the subject estimates that he has over his environment. This paper shows how this judgment of actual competence influences the processes and forms of thinking.

INTRODUCTION

It is my intention to present a hypothesis about the embedment of human thinking in a general organizational schema of behavior. Human behavior and the behavior of animals developed to guarantee continued existence of the individual and of the species. To achieve that purpose the structure of behavior must meet the requirements of individual motives as well as the requirements of the ecological conditions. In studying the role of thinking in a general organizational schema of behavior, we

*I would like to dedicate this paper to Professor H. Wegener, University of Kiel, on his 60th birthday.

are studying the ecological embedment of thinking as well. I shall proceed in the following way: First, I shall present a thesis about fundamental forms of behavior and their interactions. I shall discuss the different fundamental forms of behavior in more detail later, stressing the discussion of thinking and its function. The background of my presentation is a long series of studies and experiments dealing with human thinking in different situations. I do not intend to enter into the particulars of these studies; my aim is to achieve a significant integration.

FUNDAMENTAL FORMS OF BEHAVIOR

Talking about fundamental forms of behavior, one usually thinks of categories like "breeding behavior," "aggression," "sexual behavior," "exploratory behavior," "dominance," or "submission," etc. We shall not be concerned with such categories here, which refer to different motivations for behavior. All such categories of behavior we shall call effect-directed behavior. All forms of effect-directed behavior strive for a change of the environment or for a change of the internal state of an organism to meet the requirements of a particular motivation.

Most behavior is effect-directed and draws the attention of an organism to those parts of the environment which seem to be relevant for the actual motive. This means that effect-directed behavior divides the environment into a foreground and a background. However, since it is normally dangerous for an organism to neglect many parts of the environment, background-monitoring is necessary. I believe that it is also reasonable to distinguish two other forms of behavior besides effect-directed behavior, namely, a) orientation behavior and b) monitoring behavior.

The function of orientation behavior is to build up a representation of the current environmental background, its immediate past, and the ways in which it is tending to change.

The function of monitoring behavior is to determine whether or not the hypotheses about the development of the momentarily given situation are correct and to estimate the "uncertainty points" of the "image of the background" (see below).

The aim of effect-directed behavior is to fulfill the requirements of a current motivation. Effect-directed behavior strives for an effect on the given situation; the organism wishes to alter the current situation. The function of orientation and monitoring behavior is to secure degrees of freedom for action; beyond this they have no function. Effect-directed behavior refers to a present or anticipated need, to hunger, thirst, sexual, or informational needs. Orientation and monitoring behavior do not refer to a certain goal; they serve the function of making appropriate action possible.

These forms of behavior and their interactions establish challenges for thinking and trigger its process as we shall see below.

THE INTERACTION BETWEEN FUNDAMENTAL FORMS OF BEHAVIOR

In Fig. 1 a hypothesis about the coordination of the behavior forms is depicted. The center is effect-directed behavior. Usually an individual has a certain goal which is the result of his present motivational state. The individual is searching for something to eat, to drink, for information, for a partner, for diversion, or some combination of these.

Effect-directed behavior leads to new situations (by influencing the surroundings or by locomotion). A change of surroundings leads "automatically" to an orientation behavior introduced by an orientation reaction in the sense of the "What-is-that-reaction" by Pavlov (6). We shall describe the automatic aspects of orientation behavior later in more detail. Roughly, it consists of attempts to relate the new situation to known patterns. If this does not succeed, an exploration-motive comes into existence, which may then become the actual motive of effect-directed behavior.

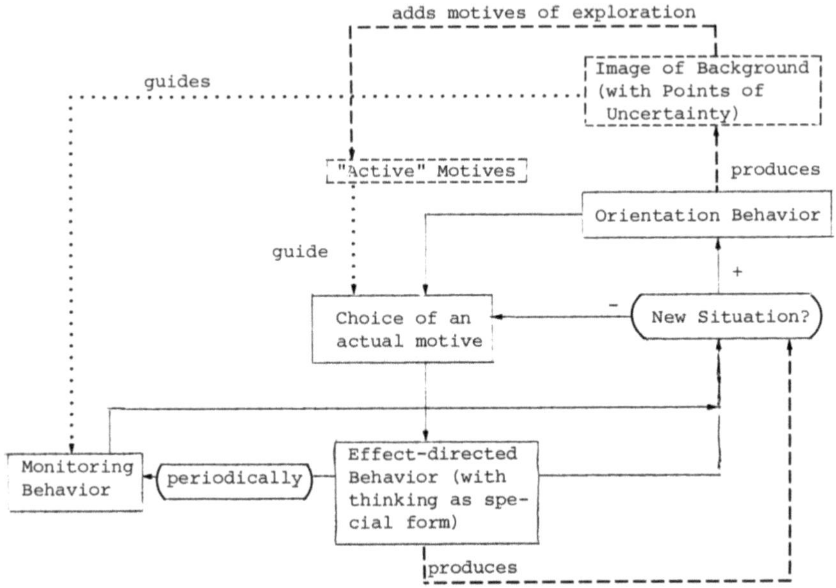

FIG. 1 - Coordination of effect-directed, monitoring, and orientation behavior. ——: sequence of information processing, ---: produces data structures, ...: influences information processing.

Effect-directed behavior is interrupted by monitoring behavior more or less periodically. This is necessary and reasonable as effect-directed behavior is confined to a specific part of the whole environment according to the specific motivation at work. To guarantee that the organism misses nothing in the environmental "background" which may be important though irrelevant for the current motivation, effect-directed behavior is sometimes interrupted by a kind of "looking about," that is, by monitoring behavior. In case monitoring behavior discloses a new situation, orientation behavior follows.

I shall now discuss the particulars of the inner structure making up these three forms of behavior.

ORIENTATION BEHAVIOR

It is the purpose of orientation behavior to form a cognitive unit, which we shall call "image of the background" (IB). This is an image of the contemporary environment, its history, and its tendencies to change. The IB is the basis for the possibility of rapidly changing the focus of attention and the course of action if necessary. You enter an unknown room and look around, noting - more or less precisely - the things which exist in this room, the position of the furniture, the light switches, doors, windows, etc. If it should now happen that you are suddenly forced to act without your action being supported by perception (suddenly light has been turned off, or a big dog rises in one corner of the room and approaches growling), then the IB permits action appropriate to the situation. You "remember" the matchbox on the couch table and can find the approximate direction in spite of the darkness. In the case of the dangerous dog, you remember the approximate position of the door, and you can move backwards slowly without losing sight of the threatening beast.

The IB may be considered as consisting of a sequence of spatial images of the surroundings stretching from the past to the future. The past part of this sequence represents the experiences of the actor through time. It is the product of a sequence of orientation behaviors in the past. The past part of the IB renders possible a backward orientation. This means, for instance, that the exit out of the labyrinth of a large office can be found. If this does not always work well, the reason is in individual- or situation-specific defects of orientation behavior. Someone deeply absorbed in thought, for instance, gives up much of his orientation behavior and will "not know where he is" if he is suddenly jerked out of his thoughts. The threshold of the orientation behavior is dependent on the intensity of the particular effect-directed behavior. Someone who acts for a very urgent and important reason will have a high threshold of orientation behavior. This can lead to disturbances of orientation or to a complete loss of orientation.

How is the IB constructed? Orientation behavior consists of a large, unpurposive reception of information. Things in the environment are sorted into categories and their spatial relations to each other are recorded. There is also a constant evaluation as to whether the whole situation fits into a general schema. Someone who enters an unknown room notes chairs, tables, cupboards, etc., and their spatial relations. If the room is furnished as rooms usually are, nothing follows. But if the set of armchairs faces the wall, if the table stands upside down, something is "wrong" and the actor may turn to exploratory behavior. We call those portions of the environment which do not fit into a general schema "points of (categorical) uncertainty." If points of uncertainty are important enough, they may stimulate further exploratory behavior.

Besides the attempt to categorize the surroundings according to existing memory schemes, orientation behavior also consists of an extrapolation process which anticipates the future part of the IB. Usually the future states of the elements of the environment are more or less easy to extrapolate by considering their present state. So it is almost certain that the walls of a coffeehouse will stay in place over a longer period of time. But who will come through a door, and when, is not at all certain! That means that there are not only uncertainties of categorization but also such of extrapolation. The points of categorical and extrapolative uncertainty are those positions of the environment which are not within the sphere of influence of the individual. The extent of possible influences on the environment is an important determinant of behavior with which we shall still be occupied.

The automatic extrapolation process which anticipates the future part of the IB shows two important process characteristics: the extrapolation takes place categorically and linearly. What does that mean?

People tend to look at future as a kind of swollen or shrunken copy of present time. For its description they use the same categories with the same weight as for the description of the present; this we call categorical extrapolation.

J. Verne could apparently only imagine a journey to the moon accomplished by a giant gunshot. Though rockets were known in his day, he was inspired to use just the gun for the purpose of spacetravel because of its predominance on the battlegrounds.

The tendency towards categorical extrapolation often received political importance. For instance, most military planners imagined World War I as a kind of extended version of 1870/71, with the engagement of horsemen and with bayonet charges accompanied by drum and whistle. This was an effect of categorical extrapolation and of great importance to political decision making.

What we mean be linear extrapolation is that people usually look at processes of change as if the rates of change were constant over time. Such an intuitive rating behavior fails to do justice to the characteristics of exponential developments as they are frequently found in biological and economical processes of growth and decline. This is the reason for serious rating mistakes (see (1,8)).

MONITORING BEHAVIOR

Orientation behavior consists of a diffuse, undirected intake of information. Monitoring behavior is directed and relates to those parts of the surroundings which allow one to expect notable events or changes, i.e., dangerous events. Monitoring behavior is necessarily related to an IB. It serves to gather information about especially crucial portions of the environment - the points of uncertainty mentioned above. The process of monitoring behavior consists of testing whether or not the points of uncertainty are still within "a setpoint region." The sniffing roe tests whether "the air is clean," i.e., whether the composition of the air is such that nothing indicates a danger. Monitoring behavior is also observed in human beings. The reader of a newspaper in a coffeehouse who drops the newspaper to check the entrance area shows monitoring behavior, as does someone who waits for the bus and sometimes looks around to see whether it is approaching. Monitoring behavior is regulated by the points of uncertainty, which determine what is observed as well as the extent of this form of behavior. The visitor to the coffeehouse monitors just the entrance area because this is a point of high extrapolative uncertainty. In the same way a conspicuously dressed visitor more often causes monitoring attention than a "normally" dressed visitor. The showy visitor is a point of high categorical uncertainty. (High categorical uncertainty implies high extrapolative uncertainty. If you do not know how to classify something, you also do not know how it will behave in the future. But there can also be high extrapolative uncertainty combined with small categorical uncertainty. The entrance area of the coffeehouse is a good example of this.)

Not only the direction but also the extent of the monitoring behavior are dependent on the points of uncertainty. If an individual feels quite unsure in his surroundings, not knowing what will happen next at many points, he will show monitoring behavior more often, and over a broader range of directions, than an individual who feels quite secure.

The weight of the momentarily pursued intention is also important for the regulation of monitoring behavior. Someone who is occupied exclusively with an important and urgent matter will exhibit less monitoring behavior than someone who is not engaged in such effect-directed behavior.

EFFECT-DIRECTED BEHAVIOR

The most complicated fundamental form of behavior is effect-directed behavior, which includes several different types. Orientation and monitoring behavior are after all "unpurposive." They do not strive for changes in external or internal states, but they are "prophylactic" forms of behavior; care is taken to be prepared for a wide range of possible situations and events.

Effect-directed behavior tends to change states of either the external or inner world. Its aim may be to change the external world in such a way that it supplies something to eat, to drink, a sexual partner, a diversion, or information. One can categorize effect-directed behavior in many ways: In the following sections we will deal with two such categorizations, namely, the different forms and the different standards of effect-directed behavior.

FORMS OF EFFECT-DIRECTED BEHAVIOR

Effect-directed behavior occurs in three forms of increasing complexity, which normally occur in the following order: a) automatic behavior, b) search behavior, c) heuristic behavior. There are no sharp boundaries between these forms of behavior; they are accents within a continuum. By automatic behavior we mean a sequence of sensorimotor coordinations which lead from the given situation to the aspired goal. In human beings such automatisms are mostly learned. I am sitting at my desk, and aware that I am hungry, I get up, go to the kitchen, fetch a piece of bread out of the breadbox, spread butter on it, take a piece of sausage out of the refrigerator, and go back to my desk. All this happens automatically, and often I am not conscious of it before it is done. The behavior has become totally automatic as a sequence of receptions, expectations, and motoric reactions. An obvious and well-known example of behavior consisting of learned automatisms is driving a car.

Search behavior as the next step of effect-directed behavior starts after failure of an automatism. It consists of searching for a stimulus which fits into the entrance scheme of a goal-directed sequence of automatisms. The hungry person who looks for the sign of a bakery or of a restaurant is an example of such a search behavior.

The last and highest level of effect-directed behavior consists in the use of heuristic techniques. (The word "behavior" is used in this case in a very metaphorical sense.) Using heuristic techniques incorporates thinking processes. This level is only within reach for animals in its simpler forms, for instance, as overt trial and error behavior. If one wants to say quite generally what is meant by "heuristic technique," one could say that a heuristic technique is a construction rule; this means a sequence of methodically used information processing steps which produce a certain cognitive configuration. Modern psychology of problem solving has identified many forms of thinking as construction processes for sequences of action units. If someone thinks about how to repair his defective car, the thinking process consists of producing sequences of manipulations at the car's motor. The thinking process is successful if an appropriate sequence leading to a goal has been employed.

The most simple form of such a construction process is trial and error behavior, which is found in many animals. One tries more or less systematically single action units and tests whether they lead to the goal or effect at least an approximation of it. Trial and error behavior a) may happen "internally" in the form of mental trials or "externally" in the form of real manipulations, b) is probably the phylogenetic source of thinking ("Thinking is internally acting on trial," Freud says) and at the same time its ultima ratio, and c) in its internalized form is the most simple way of thinking. We have also described more complicated forms of such construction processes elsewhere (2). Such complicated forms are always composed of heuristic rules which limit the choice of action units.

By the use of restriction rules, the total construction process becomes more economical as well as usually more successful. One crucial restriction rule for the choice of action units is the comparison of the given situation and the desired goal, the setting up of a list of differences between situation and goal, and the choice of action units according to this list of differences. If I am in Saarbrücken and want to go to Nürnberg, situation and goal differ in respect to the geographical position. One difference between the two cities is that Nürnberg is farther east than Saarbrücken. To remove the difference, operators have to bring me to the east, and towards Nürnberg. With this difference as a clue, one must look at those locomotion operators which can move one to the east. I need not study the flight times of airplanes from Saarbrücken to Paris, if I want to go to Nürnberg. (Perhaps all this sounds rather unimportant and obvious, but the substitution of a nondirected trial and error behavior, in which action units are arranged side by side, by a restricted and directed choice behavior is a very significant discovery.)

Thinking has developed from overt behavior by internalizing behavioral units. The internalization of behavior has two prerequisites. One can consider overt behavior of man and animal as a process of the following form: stimuli activate sensory schemata, which leads on the one hand to reactions and on the other to certain expectancies concerning the effect of reactions. The left part of Fig. 2 shows this organization schematically. An internal behavior cannot be managed with such a system, since one would have to decouple the primary stimulus receiving systems (the sense organs) and the primary reaction systems from the secondary sensory and motoric schemes. This decoupling is necessary because without it every activation of a motor scheme on trial would lead to real performance of an operation. Besides, the reception of external stimuli would always interfere with the activation of sensory schemata while acting internally on trial.

The technical trick of decoupling, necessary to allow thinking to take place as "internalized behavior," proved to be a philosophically powerful concept. This decoupling is one of the

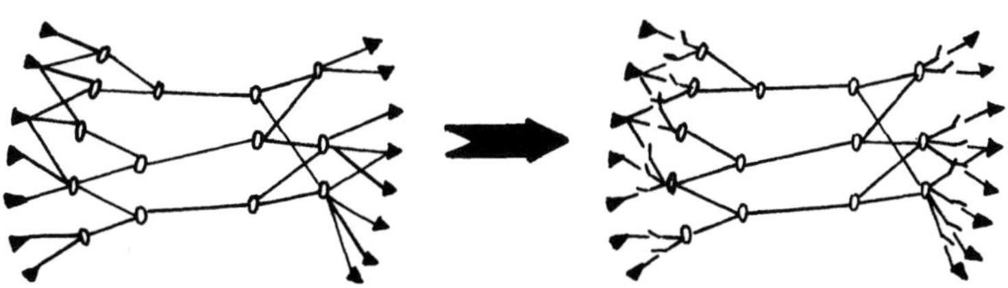

FIG. 2 - Coupled and decoupled system with a sensory and a motoric surface. The arrows indicate the general flow of direction of activation, the circles indicate memory schemata.

preconditions for recognizing a "mind-body problem." The decoupling led to the separation of an (inner) soul from a (external) world and so to a separation of a (inner) world of ideas from a (external) world of materialistic things. The "duplication" of the world into a materialistically real, objective part, and a psychologically represented, subjective portion (4) has been of greatest importance for the history of philosophy.

We regard thinking as a construction process which works on a memory system of sensory and motoric schemata decoupled from the external world, and which changes these schemata at the same time. The process consists of a systematic use of heuristic rules to restrict the search area. Quite a crucial heuristic operation is, for instance, comparing. There are many different operations which space does not permit me to describe in detail here.

The decoupling of the sensory and motor "surface" from the memory schemata is a necessary condition for autonomous thinking, which is independent of the actual situation. Another necessary assumption has to be made about the possibilities of controlling the processes of activation of the sensory and motoric schemes in the decoupled state. In normal behavior, motivation, perception, and expectation control the processes of activation.

A certain motivation produces the imagination of a goal. The perception of the actual situation together with the anticipation of the goal produce associations concerning the discrepancies between the momentarily given situation and the goal. The hungry bird, starting a search for food, has a search image which is activated by motivation. In his search for food the bird uses the perception of the situation in order to associate appropriate actions with the motivation and to monitor their outcomes.

In the decoupled state of internal simulation no perception of the situation exists. The only possible starting point for associations is motivation. Within a network of sensorimotoric schemata, motivation alone could produce only associations concerning possible goals of the current motivation. The resulting disorder may be present in dreaming and daydreaming in humans, perhaps also even in some animals. This kind of thinking is not a very effective method for constructing an internal connection between the present situation and the goal, since it lacks direction. It would be advantageous to have a control system which regulates the otherwise disorderly thoughts. This system in human beings is language, a "second signal system" (6) which replaces the role of perception. A sentence, as a sequence of words bound together by grammatical rules, is, technically speaking, a device for organizing these disorderly thoughts. The rules of language aid in this organization. In this way language allows an ordered flow of thinking, which replaces the disorderly sequence of associations of dream-thinking.

After having decoupled the internal and the external world, it is necessary to designate a new system for the activation of "imaginations." The concept "car" makes the activation of the sensory scheme "car" independent of the existence of a real car. Language permits the heuristic system to utilize the elements of memory independent of stimuli from the external world. Therefore the most important parts of the human heuristic system are the rules of language production.

It it is true that apes are not able to combine the words of a (ape) language in meaningful ways, utilizing grammatical rules as Ristau and Robbins (this volume) point out, this might indicate a severe restriction to the ape's thinking abilities. Having no access to such a construction plan means that the sets of images, which are associated with the words, cannot be brought into a clear combination and remain isolated, without forming a complex structure.

Autonomy of internal behavior requires the ability of the individual to distinguish between those images activated by perception and those images activated by the second signal system. This means that the individual must be able to distinguish between activities which are forced upon him by the external world and those activities which are generated by his heuristic system. Autonomy requires self consciousness as a prerequisite.

The autonomization of the "internal behavior" by decoupling the sensory and motoric surface from the internal schemes and by establishing a secondary, language-led control system brings great possibilities for improving the effectiveness of purposive behavior, but it brings dangers as well.

On one side, an autonomous "internal behavior" renders possible the anticipation of future situations and the imagination of situations which do not exist at the moment. Autonomous "internal behavior" even renders possible the imagination of situations or actions which are personally, momentarily, or completely impossible. This greatly enlarges the possibilities of planning the future and of acting. On the other hand, autonomy of thinking provides the danger of autistic degeneration by habitualizing the decoupling of the internal and the external world. This is the danger of a retreat to phantasy and the world of imagination, which may be much easier to handle than the real world.

EFFECT-DIRECTED BEHAVIOR AND COMPETENCE

Effect-directed behavior means actively seeking a desired state or trying to avoid an undesired state, and as mentioned above, it takes place in the following ascending order of complexity: automatic behavior, search behavior, and heuristic behavior. This sequence is also temporal and its extent and nature are dependent on the perceived competence of the individual. By perceived competence we mean the extent of control over his surroundings the individual believes that he possesses. The taxation of the extent of control is connected with orientation and monitoring behavior and is proportional to the number of points of uncertainty in the IB (see above). The larger the number of points of uncertainty, the smaller is the possibility of control, the smaller the perceived competence. Inability to control, however, means unspecifiable danger. The danger of being run over by an oncoming car is a specifiable danger. One can calculate what may happen. But the danger coming from points of uncertainty is not specifiable because what may happen cannot be foreseen. To characterize the differences between specific and unspecific danger, language discriminates between fear and anxiety. Fear arises with the anticipation of a specifiable danger; anxiety characterizes the state of feeling an undefined, inconceivable form of danger.

What is the most adequate reaction to an unspecifiable danger? We assume that two postulations have to be fulfilled:

1. A state of high readiness to act has to be maintained, i.e., the sensory and motor thresholds should be low.
2. The forms of action for which one prepares oneself must be unspecific, i.e., they must be suitable for all possible threats, not only for a specific one. Such unspecific forms of action to overcome possible threats quickly are, e.g., flight and aggression. Whether in a situation of danger either flight or aggression is chosen depends on the distance from the threatening object. Being further away, one will try to avoid; being close, one will tend more towards aggression, as avoiding becomes more hopeless.

In an extreme situation, the combination of these two conditions leads to a high readiness to unreflected primitive reactions. This means that an elimination of heuristic behavior will take place. A high state of readiness to act is incompatible with a high extent of decoupling the internal world from the external, which thinking postulates. Even if it is only in extreme situations that loss of control leads to a total elimination of reflection, unspecifiable dangers must, nevertheless, project these simplified elements of behavior into the thinking process.

To guarantee a quick readiness to act, the thinking process has to be simplified. If the internal simulation of actions is abbreviated, it is impossible to consider many conditions and secondary and long range effects of actions. Thinking relates only to the essence, the background is eliminated. Thus judgements and decisions as products of thinking processes become more extreme and risky. This restriction of the thinking process, called an "intellectual emergency reaction," is described in another paper (3).

The tendency to run into forms of black-white thinking in critical situations (7) was demonstrated after studying verbal behavior of political leaders in crisis situations. The "intellectual emergency reaction" which allows the calculation of conditions and consequences of actions only in an abbreviated form is an extremely explosive characteristic of the "adaptations" of human thinking to ecological conditions. It is just in those situations when it would be of importance to anticipate many conditions and consequences by "internal simulation" that human beings tend to abbreviate and isolate thinking.

There is not much comfort in the statement that this reaction was originally a biologically reasonable one. It guaranteed the stimulus-reaction-readiness to act, the immediate striking or retreating. If actions consist not of motor reactions, but of mainly passing on information by pressing a key, by telephoning, teletyping, etc., the switch-over from "intellect to muscle," included in the intellectual emergency reaction, is no longer reasonable.

Flight and aggressive tendencies mingle in the thinking process long before they come forward as real acts. Thus flight reaction in thinking tendencies may cause one to occupy oneself with unimportant but well-mastered activities instead of the urgent and important problems. Who has not caught himself preferring to clean out his desk drawer or to rearrange his library rather than coming to grips with a complicated problem?

In crisis situations Louis XVI occupied himself for days with art metal work instead of the urgent economic problems of his country (5). Here a flight reaction is suggested. The change from topic to topic observed in uncertain and complex situations, the inconstancy of thinking which does not succeed in completing a single activity, is a flight tendency which renders possible the withdrawal of oneself from pressing realities (described in more detail elsewhere (3)).

Aggressive tendencies in thinking are also found in situations of low control prior to aggressive actions. One impressive example may suffice: In one of our experiments, after fruitless effort to raise the productivity of an (computer simulated) industrial enterprise, one subject came to the decision to shoot workers whose machines showed deficiency "because of sabotage." This subject later claimed this decision to be a joke, but it seems to show the direction of thinking in such situations of helplessness.

CONCLUDING REMARKS

We tried to describe the embedment of human thinking in the whole system of human behavior organization. It was our main interest to illustrate the role of thinking in the control system of an organism. Thinking is not an isolated psychological act, it takes place and assumes specific forms according to the conditions of the situation. These dependencies of thinking, however, are not laws comparable, for instance, to the laws of physics. Thinking can adapt itself and its procedures to its own objectives and thus achieve some degree of independence. But this happens unfortunately seldom.

REFERENCES

(1) Bürkle, A. 1979. Eine Untersuchung über die Fähigkeit exponentielle Entwicklungen zu schätzen. Thesis, FB 06, University of Gießen, W. Germany.

(2) Dörner, D. 1979. Problemlösen als Informationsverarbeitung. Stuttgart: Kohlhammer.

(3) Dörner, D. 1980. On the difficulties people have in dealing with complexity. Simul. Games 11: 87-106.

(4) Graumann, C.F. 1965. Bewußtsein and Bewußtheit Probleme und Befunde der psychologischen Bewußtseinsforschung. In Allgemeine Psychologie I. Der Aufbau des Erkennens 1. Halbband: Wahrnehmung und Bewußtsein, ed. W. Metzger. Göttingen: Hogrefe.

(5) Kühle, H.J. 1981. Handeln in der Geschichte - Eine Fallstudie über Ludwig XVI von Frankreich. Manuscript, University of Bamberg.

(6) Pavlov, I. 1974. Die bedingten Reflexe. Munich: Kindler.

(7) Suedfeld, P. 1978. Integrative Komplexität als eine Variable historischer Forschung und internationaler Beziehungen. In Kognitive Komplexität, eds. H. Mandl and G.L. Huber. Göttingen: Hogrefe.

(8) Wagenaar, W., and Sagaria, S. 1975. Misperception of exponential growth. Percept. Psychophys. 18: 416-422.

Animal Mind - Human Mind, ed. D.R. Griffin, pp. 113-130.
Dahlem Konferenzen 1982. Berlin, Heidelberg, New York: Springer-Verlag.

Social Knowledge in Free-ranging Primates

H. Kummer
Ethology and Wildlife Research, Institute of Zoology
University of Zurich, 8050 Zurich, Switzerland

Abstract. Assessing an animal's knowledge requires that the researcher knows the animal's goals and means at least as well as the animal does. This is more difficult with regard to the animal's knowledge about his social companions than about extra-specific objects. While social knowledge as evident in the natural habitat is in some cases accessible to simple experiments, social intelligence requires long-term laboratory studies. The field examples reported in this paper confirm that apparently simple behavior such as "hiding" or "not fighting" may involve more knowledge about others than complex interactions like "protected threat"; parsimonious behavior may result from unparsimonious internal processes.

INTRODUCTION

Looking back over the past 25 years, I am impressed by the fact that research on nonhuman primates has provoked increasingly demanding hypotheses on their mental capacities, including their assessing each other as "social tools." Yet, cognition and social intelligence have not been a formal topic of my research. Thus, the present paper describes the first steps of a primate ethologist toward the topic of cognition.

The new "Cognitive Ethology" seems to aim at something that already has a long tradition in psychology (18). The main ethological contribution I can see at present is to describe natural situations in which cognitive abilities are most apparent and which possibly represent the functions for which these abilities were evolved.

The term cognition is alien to traditional ethological thinking. It seems to suggest something about the nature or quality of the representations of the world in an animal. To a biologist, the morphology of a parasite is a representation of the host, an innate motor pattern a representation of its environmental object, although both are, in a sense, negative images. Cognitive processes seem to involve decision making that uses environmental representations in the brain. While this seems clear enough, some psychologists may not realize how difficult it is for a traditional ethologist to understand precisely what types of images qualify as "representations." Both "cognition" and "representation" seem to resist operationalization in animals. The conference itself has not changed this impression.

I shall use the everyday term "knowledge" to refer to an internal state or process that helps to carry the animal from a start to a goal. By "goal" I mean a state which the animal perceives and which regulates its behavior until the goal is reached, regardless of whether that state is adaptive in evolutionary terms. Such "knowledge" may be more or less appropriate or "correct" according to the assessment by the observer. The richness of knowledge in an environmental field is revealed by the number, complexity, and optimality of observable pathways between all possible starts and goals in a context. Innovative intelligence can be estimated by the difference between the most parsimonious path chosen in the present situation and the most similar path which the subject has used often and successfully enough to learn it during his lifetime (which corrects for his adaptation by learning) and which his ancestors have used often and successfully enough for a genetic program to be formed (which corrects for his adaptation by evolution). This crude description is based on the idea of an environmental maze. In order to infer the minimal knowledge required to run a set of pathways, the observer must himself know the start and the goal, and his knowledge about the number of pathways and their appropriateness to the goal must at least equal the knowledge of the animal. This is where our problems begin.

The traditional laboratory psychologist makes sure by manipulation that the goal is eating or drinking (which he can see), that he can place the animal into the start position he prefers, and that he perceives and understands all possible pathways that are open to the animal (which he may achieve by taking a bird's-eye view of the maze). He also makes sure that the animal knows nothing more about the environmental situation than he does.

Von Frisch, experimenting with his bees, was in nearly such a position. The bees had a single goal and a fixed start, both of which he knew, and the correct path was generally a straight flight. Orientation and the use of inanimate tools are the "natural" behavioral systems in which the requirement of the observer's superior knowledge about goals and pathways is most easily met. By contrast, a field primatologist sitting ten feet from a hamadryas baboon who is gazing out into the desert and about to depart for the day's foraging trip knows only that the baboon has several different goals (at least food, water, security, and not losing the clan). He knows far less about plant sites, food toxicity, the state of waterholes, and the routine of predators than the baboon does. The observer was not around when the baboon learned his topography, nor when the baboon's ancestors evolved their abilities to learn it. In addition, the relevant objects of knowledge change over time, and both the baboon and his observer may therefore be mistaken in their assessment. Even so, the skilled ecologist is able, at least potentially, to see most of what his baboon sees when they survey together a food site or another object of the extraspecific environment.

My topic in this paper concerns a primate's knowledge about the members of its group, particularly about the qualities, tendencies, and the availability of each individual companion with regard to its own goals (15). The observer's difficulties here in knowing start, paths, and goal are even more aggravating. Orientation in space, finding a certain fruit, or using a stick are rather general problems familiar to many species and certainly to the human observer; assessing the social qualities and changing tendencies of an individual baboon is a task for which only baboons have evolved specialized skills. It is specific to the species. Thus, primatologists have spent years in trying to learn themselves the "social maze" in which their subjects perform. The social maze may be difficult even for the baboon, since the motivational changes of his companions are certainly faster than changes in his habitat, predators excluded. Even so, one of my first observations in the field (11,12) was that a hamadryas baboon female uses her male as a "social tool" in the so-called protected threat. By presenting her rear to him and at the same time threatening another female she sometimes succeeds in inducing the male to attack her opponent.

Since baboons do not use mechanical tools, it appears that primates evolved the use of social tools long before they used sticks and stones (3,13). Humphrey (9) actually stated that "the chief function of intellect is to hold society together," and he may have implied that this is also the most ancient function. At closer inspection, the relative antiquity of social tool use is surprising for while a maze or a simple mechanical tool remains the same throughout the episode of its use, a "social tool" changes (its motivations) and follows its own goals. It is a maze that responds, a hammer that hits back or accelerates its movement. Second, many qualities of a mechanical maze or tool are open for the senses to survey, whereas many relevant qualities of the social tool are hidden in its brain.

One will correctly object, however, that a primate companion often signals his intentions, and that the breaking point of a stick may therefore be more difficult to predict than a companion's attack. Nevertheless, I still believe that a living primate is a more difficult tool. That social tool use occurs in many primate species, whereas only the great apes have been observed to routinely use mechanical tools in the wild could be explained speculatively as follows: 1) Social life as such was a long-standing preadaptation to social tool use. 2) The social tool was of more importance to reproductive survival than mechanical tool use, its benefits selecting for specific learning programs in the social field. Such learning programs would make social tool use an adapted rather than an adaptable behavior, and at least partly prewired rather than creatively intelligent. 3) A primate has nearly permanent opportunity for social learning. 4) The prediction of conspecifics might be facilitated by empathy among the conspecifics.

Each of these explanations, if it is true, sets the human observer at a disadvantage, since he lacks the intraspecific adaptations. The field worker cannot be certain about the short- or long-term goals of a socially interacting primate, and his knowledge about the alternative pathways toward those goals is mostly fragmentary guesswork. And whereas the experimenter can manipulate a maze and a mechanical tool, it is difficult to manipulate the behavior of a stimulus companion under field conditions. (However, Sigg (22) has recently succeeded in altering the spatial formation of natural hamadryas baboons in an enclosure by conditioning.)

With these difficulties it is not surprising that my search of the literature has not yielded any field studies on social intelligence or closely related topics on nonhuman primates. Every field worker has of course seen impressive behavioral sequences where he was confident that he recognized the short-term goal of a monkey or ape and that he knew something about the pathways available to the animal. However, such rare stories were traditionally omitted in scientific papers as mere "anecdotes," and anecdotes in the literal sense they therefore remained. This has been unfortunate. As a consequence, every field worker knows only the anecdotes recorded by his research group, and he can judge with confidence only what he has seen himself. The beginning of systematic field studies is, however, apparent (21).

A good anecdote describes an event in which the particular situation strongly favors a single interpretation, for example, when a rare behavior coincides with a rare external circumstance to which it seems adaptive, particularly if other plausible causes and functions are absent. The following observations by A. Kern (personal communication) illustrates this: A male baboon sits at the upper edge of a cliff. As he does so, he loosens a stone. He immediately grabs it and holds it for a few minutes. Directly beneath him a group of juveniles are playing. Just after they have moved away, the male baboon drops the stone, and it falls on the deserted playground. Since baboons very rarely sit with a stone in their hand, the male might just possibly have held the stone in order not to hurt the juveniles. The anecdote thus provides the idea for a series of experiments.

In the following sections I will present some interesting types of social behavior which I have seen myself or about which I could thoroughly question a colleague. Some of these anecdotes suggest research questions or experiments on what the animals may know about each other.

HIDING

Typical field episode: A juvenile female hamadryas baboon in estrus leaves her adult male leader and repeatedly mates with a juvenile male behind a rock where her leader cannot see her. Between matings, she goes to where she can peek at the leader, or even approaches him and presents herself to him before she again mates with the juvenile in the hiding place.

The assumed goal of the female is mating. The following explanations of her hiding behavior are ordered according to increasing complexity of knowledge required from the female.

Explanation 1: The female is in conflict as to which of the two males she prefers for mating, and she uses the rock in order to "cut-off" (4) the visual stimuli from the leader. The explanation requires no particular social knowledge. It is contradicted by the female's looking for the leader between matings and by the fact that leaders are generally reluctant to mate with juvenile females.

Explanation 2: The female knows a) that the leader may attack her when she mates with a juvenile male (which is correct according to the observer's assessment). She also knows b) that the attack is unlikely when she mates in a particular place, e.g., on the south side of a particular rock (which is incorrect). This explanation is contradicted by the observation that she generally selects the side of this or any rock opposite the momentary place of the leader.

Explanation 3: The female knows a as above. But instead of b, she knows c) that the attack from the leader is unlikely when she does not see him (which is more correct than b but not entirely correct). The explanation is contradicted if the female does not simply close her eyes or if she often chooses a spot from where she can see the leader but cannot be seen by him, e.g., a spot closely behind thick brush.

Explanation 4: The female knows a; in addition she knows d) that the attack is unlikely if the leader cannot see her even though she sees him (this is the correct knowledge in the observer's assessment). The following sequence suggests that knowledge d may be available: A female spent 20 minutes edging herself into a sitting position where a rock hid her front and arms from her leader while allowing him to see the top of her head and her back. She thus groomed a subadult male, an activity which is often not tolerated by leaders. The leader was able to see that she was present, but not what she was doing.

Anecdotal evidence suggests also that explanation 4 may be correct for "acoustic hiding." During an experiment that tested the "loyalty" of female gelada baboons (14), each female was confined with a colony male other than her mate while the latter was kept behind a concrete wall but within ear-shot of the confined pair. All three participants had had ample opportunity to learn their bonding pattern. When females began to groom the "wrong" males, the latter would suppress the loud vocalization which males normally utter at this moment. The males' assessment of the situation seemed to agree with that of the experimenter.

The knowledge involved in hiding tactics could be made easily accessible by experiments using various visual and acoustic screens. Such techniques are now used by Anzenberger of our group in the analysis of primate social relationships.

Similar questions can be asked about the reverse situation: Does a subject know whether and when a companion who is at present invisible will appear again? Male hamadryas baboons in the wild do not attempt to take over the female belonging to another male of their band when the latter happens to be out of sight, even when the female is close and her owner remains invisible for up to half an hour. However, if a leader is trapped in a fully transparent cage with his females outside, some band males appropriate his females within minutes, as we found in a series of field experiments (17). The trapped owner, even though he is close and in full view, is a less effective inhibitive agent than the invisible, absent owner, whose relationship with the female is "merely" known to others.

LINEAR DOMINANCE ORDERS

Some dyadic interactions among free-living old world primates are consistently asymmetric in that A exclusively or more often avoids B than vice versa (e.g., (8)). The dyadic asymmetries of a group can very often be arranged in a linear "dominance" order, so that if A>B and B>C, the asymmetry among A and C is A>C. Circular triads disappear within days after a group has been newly established in captivity. The linearity phenomenon is perhaps less trivial than it seems. A simple explanation is that group members rank each other on the basis of one or several variables (e.g., size or age) in which each individual occupies a point on the scale. Individuals would have to agree on which variables are to be summed in what way in order to reach linearity.

For a parsimonious explanation it is sufficient that only one individual in each dyad compare the two individual values. Since in a contest it is the loser who switches his behavior from fighting to submission or escape, he is the more likely comparator, while the superior animal may learn that he is superior only from that change in the behavior of the inferior. Rowell (20) pointed out that dominance can be recognized primarily from the behavior of the inferior, and de Waal (25), in captive groups of Java monkeys, found better linearity for submissive than for aggressive behavior, which supports this explanation. In a recent review, Deag (6) qualified Rowell's conclusion. He found higher correlations between the frequency order of agonistic behavior and rank for the dominant role than for the subordinate role. The dominant animals in his group of Barbary macaques certainly contributed to rank order.

A second explanation requiring that <u>both</u> members of a dyad make the comparative judgement implies a rather demanding hypothesis. Linearity will emerge only if an individual's assessment of its own value on the scale agrees with the assessment of that value by other group members. In a linear triad $A>B>C$, B judges himself to rank lower than A and higher than C; his implied judgement that $C<A$ obviously agrees with that judgement by C and A themselves. I am not aware of evidence that rejects the simpler explanation (that the first and decisive comparison is made by only one pair member) for nonhuman primates, but some available dominance matrices from newly formed captive groups indicate that they respond to the magnitude of the difference of the underlying variable (25).

Some observations suggest that a primate can learn from observing dominance interactions among two others (20). A subordinate individual would reduce the number of contests by concluding that A is superior to him if he sees B lose to A after he himself has lost to B. Experiments on how much about the dominance order a primate can infer on the basis of incomplete experience could be useful.

KNOWLEDGE ABOUT CONSEQUENCES OF BEHAVIOR IN A DYADIC RELATIONSHIP

Episode: During field work, two adult male hamadryas baboons were transported in the same 2x1x1 meter cage. During the entire transport, the males sat at opposite ends of the cage, each turning his back to the other, avoiding looking at the other and hardly moving. The episode is of particular interest since the basically aggressive males succeeded in avoiding a fight in a situation of narrow confinement that never occurs in nature. Contrary to uninformed expectation, they neither tried to escape from the cage nor kept a close watch over each other.

The assumed goal of each male is avoiding a fight. The minimal acceptable explanation seems: Each male knows that fighting at such close distance is imminent and would be very dangerous (which is correct according to the observer's assessment), that eye contact and any hasty movement greatly enhances the probability of fighting (correct), and that the companion will reliably behave as if he knew all this as well. An incorrect assessment on this last point would have severe consequences.

Similarly, laboratory experiments and field observations (16) strongly suggest that hamadryas baboons assess the risk of a fight with a <u>particular</u> companion. A dominant baboon risks a fight over a female or food seen in the possession of another particularly if they are otherwise on intimate terms. This may be so because he can more precisely predict the outcome of a fight against an intimate, or because he knows that their relationship is too well established to be damaged by a fight. Such assessment of how one's behavior may affect one's relationship with another is evident in the following observation by

J.-J. Abegglen (personal communication). When two hamadryas males are convened for the first time, they develop their relationship through a sequence of behavioral steps. At one stage, the subordinate presents his rear and the dominant male mounts him. Among males that were particularly incompatible as expressed in their low tendencies to groom, the dominant male reversed this by presenting himself and allowing the subordinate to mount. It was evident to the observers that the males' goal was to establish friendly relationships. The parsimonious explanation of the observed reversal is that the dominant male found the detour of playing inferior merely by trial and error, but this is not likely in view of the consistent use of presenting as an appeaser even by dominant males in natural situations. For example, family males present themselves to their future challengers months before the latter actually deprive them of their females (1).

Packer (19) observed that male anubis baboons assist each other in obtaining a female consort for mating. By certain gestures, one male solicits the support of another who may then join him in threatening a third male, particularly if the latter is already in consort with a receptive female. In all 6 cases where the coalition was successful, the female was obtained by the solicitor while the solicited male continued to fight the opponent. In biological terminology, the solicited male in such an episode displays altruism, which is defined as behavior that benefits another at some cost to the altruist (no moral intent is required). In the present case, the solicited male risks being wounded without a benefit, since he will not share in mating the conquered female. It is difficult to explain how such altruism could evolve unless a) the support is given only to a close relative, or b) a male primarily supports males that will in turn assist him on future occasions. The second explanation is termed "reciprocal altruism" (24). In Packer's group, each male tended to request aid from a male who in turn requested aid from him. Since the members of some reciprocating allies had been born in different troops and were most probably unrelated, reciprocal altruism is the likely explanation. This raises the question of how the individual male prevents his being "cheated" by a companion who solicits aid but never or rarely gives it. Does he remember the individual episodes and decide on whether to help or not according to how the numerical balance stands, as would a human merchant or a tribesman enumerating the services which his family has given to that of an opponent? The actual solution among nonhuman species may be less mental and more emotional (see also (24)). Primates, like other vertebrates, form dyadic bonds based on individual recognition. Establishing such a bond requires time, attention, and many social interactions. Once the bond is formed, each partner has become an investment to the other (15). The partners now behave as if they were attached to their social investment; each follows the other and defends him against rivals or even against extraspecific dangers. Future research may investigate the emotional bond as a possible premental mechanism for reciprocal altruism. Bond members aggressively prevent each other from exchanging altruistic acts with third parties; in addition, cheating by one member might reduce the emotional attachment of his partner. The cheater would thus risk losing his investment, which might easily be more costly than that of returning the altruistic acts. (That the "honest" member would also lose his investment is irrelevant to the cheater's situation. The honest member should abandon the cheater if being cheated will be more costly in the future than investing in a new bond.) It is not known to date whether cheating does reduce the attachment of the victim, but the struggle for bonding partners can be readily observed in newly formed groups. Generally, a primate does not break a bond abruptly. A slight shifting of his attention to other group members acts as a sufficient "threat" to the partner and causes that partner to interact more intensely with his bond mate.

KNOWLEDGE ABOUT THE RELATIONSHIP BETWEEN TWO OTHER ANIMALS

Infants of at least two macaque populations and probably some baboons acquire the dominance rank of their mother. This "dependent rank" (recent short review in (25)) is learned both by the infant and by his often much stronger opponents as a result of the mother's assistance given to the infant in its agonistic encounters. Mother-dependent rank in macaques is generally maintained beyond the mother's death, although Kawai (10) describes leader-dependent rank in females which requires the actual presence of the leader. These examples, if interpreted parsimoniously, do not require the subject's knowledge about the relationship between his opponent and his supporter. Even in the second example, it may simply learn that the presence of the supporter is a condition of his own superiority over a particular opponent.

In hamadryas baboons, a male rival is more likely to encroach on the female owned by another male if the latter ranks low in the female's preference as measured in choice tests (2). The experiment does not preclude the possibility that the female's preference is recognized by the rival only from signals emitted by the owner. Recognizing the relationship between owner and female would require at least that the rival responds to a response by one mate to a signal from the other. In a relevant experiment, the female would have to be visible to the owner but not to the rival.

Cheney and Seyfarth (5) played recorded screams of free-ranging juvenile vervet monkeys (Cercopithecus aethiops) to groups containing the juveniles' mothers and other adult females. Not only did the mothers respond specifically to the screams of their own juveniles, but the other females would look at the mother even before she had herself responded. This result strongly suggests that the females knew about kin relationships in their group, at least by an association of a juvenile's scream with the observed and anticipated response of its mother.

USING A COMPANION AS A SOCIAL TOOL

As I have not myself observed the well-known use of young infants by male Barbary macaques to ease the tension among them

(7), the protected threat already mentioned above will serve as an example of the behavior type. Protected threat has three effects: a) the presenting protects the actor from an attack by the male to whom she presents; b) the threats by the actor may release threats from the male against her opponent; and c) the actor's intermediate position in space prevents the opponent from approaching the male. The behavior first appears in the juvenile of 2 years in the form of sitting close to a dominant animal (usually the mother) and screeching while looking at the opponent. At 3 years, the subject presents to the dominant animal rather than sitting close. Finally, the adult subject replaces screeching by threatening and, eventually, lunging at the opponent. Females at puberty were seen to lunge too far away from the dominant male; experienced opponents used this at once to put themselves into the favorable intermediate position. The young females learned to do better within a few weeks ((11), p. 56). Most primatologists would agree that "a profitable use by the monkeys of the recruiting effect of appeal-aggression requires a good insight in the structure of social relations within the group" (25).

While such episodes are complex events, nothing about their development so far suggests intelligent invention. The observations can be explained as learning the effects of one's behavior on others and as combining these effects in skillful spatial maneuvers. The very complexity of these tripartite tactics has so far precluded experimentation.

Lindauer, in an interesting comment on this paper, pointed out that the tools used by social insects are much more often live companions than inanimate objects. For example, weaver ants glue the edges of leaves together by holding a larva in their mandibles and moving it back and forth between the leaves. By slight pressure, the worker ant causes the larva to secrete a silk thread which fastens the leaves together. Lindauer's examples indicate that in insects, too, the use of conspecifics as tools phylogenetically predated the use of inanimate tools.

LONG-TERM PREDICTION OF A COMPANION'S BEHAVIOR

When a group of primates splits into subgroups, do they know when and where they will meet again? The work by Stolba (23) suggests the possibility that clans of hamadryas baboons communicate about the direction of the waterhole where they will reassemble at noon. When the clans left the sleeping cliff in the morning, the direction of their joint march for the first 250 meters more often pointed in the direction of the waterhole to be visited that day than did the direction of later segments of the march. Clans that had separated during the morning reassembled at the same waterhole only if their joint departure had pointed in that direction. This is compatible with the hypothesis that the starting direction stands for and communicates the direction of the day's drinking place.

The example illustrates a suitable way of investigating whether a primate A knows what his companion B will do not only in the next 10 seconds, but in the next few hours. The research question requires that no communication between A and B is possible during the time span to be bridged by A's prediction.

PARSIMONIOUS EXPLANATION

Most biologists of my generation have been taught to prefer a parsimonious to a complex explanation. The recent development in primatology raises doubts about this recommendation. In part, it appears as mere opportunism: There are generally fewer parsimonious than complex hypotheses to a phenomenon, and often, one is simply unable to conceive or understand a more complex hypothesis before one has thoroughly worked with a more simple one. Furthermore, we make our objects of study appear more simple than they generally are if we systematically accept the most parsimonious explanation. This is permissible in a rapidly developing field of research where such errors are soon corrected. However, where there is little hope for new and decisive research in the near future, the simplistic bias will tend to consolidate unless acceptance of the parsimonious explanations is replaced by an awareness of the range of possible ones.

The difference between animals and humans, for example, is even now artificially widened by the principle of parsimony simply because we know less about animals. Finally, I feel uncertain as to how parsimony is to be judged. The simplicity and paucity of mechanisms required on a given level of analysis are only one measure. Introducing "higher levels" of explanation, such as intent, consciousness, or the force vitale is also considered unparsimonious, apparently for reasons other than "complexity." What exactly are these levels? Can they be operationalized? If not, what is their usefulness? The conference has revealed prospects of profitable cooperation between philosophers and empirical scientists on such questions.

Acknowledgement. This paper owes much to the participants of a seminar on Social Awareness, particularly to G. Anzenberger, V. Dasser, R. Keller, H. Moser, B. Nievergelt, E. Stammbach, and D.C. Turner. I thank J. Crook and D.R. Griffin for helpful comments on the wording of several sections.

REFERENCES

(1) Abegglen, J.-J. 1981. On Socialization in Hamadryas Baboons. East Brunswick, NJ: Associated University Press.

(2) Bachmann, C., and Kummer, H. 1980. Male assessment of female choice in hamadryas baboons. Behav. Ecol. Sociobiol. 6: 315-321.

(3) Chance, M.R.A. 1961. The nature and special features of the instinctive social bond of primates. Viking Fund Publ. Anthrop. 31: 17-33.

(4) Chance, M.R.A. 1962. An interpretation of some agonistic postures; the role of "cut-off" acts and postures. Symp. Zool. Soc. Lond. 8: 71-89.

(5) Cheney, D.L., and Seyfarth, R.M. 1980. Vocal recognition in free-ranging vervet monkeys. Anim. Behav. 28: 362-367.

(6) Deag, J.M. 1977. Aggression and submission in monkey societies. Anim. Behav. 25: 465-474.

(7) Deag, J.M., and Crook, J.H. 1971. Social behaviour and "agonistic buffering" in the wild barbary macaque Macaca sylvana L. Folia primat. 15: 183-200.

(8) Hausfater, G. 1975. Dominance and reproduction in baboons (Papio cynocephalus). Contr. to Primatology, vol. 7. Basel: S. Karger.

(9) Humphrey, N.K. 1976. The social function of intellect. In Growing Points in Ethology, eds. P.P.G. Bateson and R.A. Hinde. Cambridge: Cambridge University Press.

(10) Kawai, M. 1958. On the rank system in a natural group of Japanese Monkey, I and II. Primates 2: 111-112, 131-132.

(11) Kummer, H. 1957. Soziales Verhalten einer Mantelpavian-Gruppe. Beiheft Schweiz. Z. Psychol. 33: 1-91.

(12) Kummer, H. 1967. Tripartite relations in hamadryas baboons. In Social Communication Among Primates, ed. S.A. Altmann. Chicago: University of Chicago Press.

(13) Kummer, H. 1971. Primate Societies. Group Techniques of Ecological Adaptation. Chicago: Aldine Publ.

(14) Kummer, H. 1975. Rules of dyad and group formation among captive gelada baboons (Theropithecus gelada). In Proceedings from the Symposia of the Fifth Congress of the International Primatological Society, eds. S. Kondo, M. Kawai, A. Ehara, and S. Kawamura, pp. 129-159. Tokyo: Japan Science Press.

(15) Kummer, H. 1978. On the value of social relationships to nonhuman primates: A heuristic scheme. Biology and Social Life. Social Science Information 17. London and Beverly Hills: SAGE.

(16) Kummer, H.; Abegglen, J.-J.; Bachmann, C.; Falett, J.; and Sigg, H. 1978. Grooming relationship and object competition among hamadryas baboons. In Recent Advances in Primatology: Behaviour, eds. D.J. Chivers and J. Herbert, vol. 1, pp. 31-38. London: Academic Press.

(17) Kummer, H.; Götz, W.; and Angst, W. 1974. Triadic differentiation: An inhibitory process protecting pair bonds in baboons. Behaviour 49: 62-87.

(18) Mason, W.A. 1976. Windows on other minds. Book Review of D.R. Griffin's: The Question of Animal Awareness. Science 194: 930-931.

(19) Packer, C. 1977. Reciprocal altruism in Papio anubis. Nature 265: 441-443.

(20) Rowell, T. 1974. The concept of social dominance. Behav. Biol. 11: 131-154.

(21) Seyfarth, R.M., and Cheney, D.L. 1981. How monkeys see the world: A review of recent research on East African vervet monkeys. In Primate Communication, eds. C. Snowdon, C.H. Brown, and M. Petersen. Cambridge: Cambridge University Press.

(22) Sigg, H. 1980. Differentiation of female positions in hamadryas one-male-units. Z. Tierpsychol. 53: 265-302.

(23) Stolba, A. 1979. Entscheidungsfindung in Verbänden von Papio hamadryas. Ph.D. Thesis, University of Zürich.

(24) Trivers, R.L. 1971. The evolution of reciprocal altruism. Q. Rev. Biol. 46: 35-57.

(25) de Waal, F.B.M. 1977. The organization of agonistic relations within two captive groups of Java monkeys (Macaca fascicularis). Z. Tierpsychol. 44: 225-282.

Animal Mind - Human Mind, ed. D.R. Griffin, pp. 131-144.
Dahlem Konferenzen 1982. Berlin, Heidelberg, New York: Springer-Verlag.

Primate Social Intelligence: Contributions from the Laboratory

W. A. Mason
Psychology Dept. and California Primate Research Center
University of California, Davis, CA 95616, USA

Abstract. From a functional standpoint, wanting and knowing are the two most fundamental themes in the evolution of behavior. The evolutionary changes that have occurred in the processes subserving wanting and knowing provide the distinguishing features of intelligence and are the source of interest in intelligence as a biological phenomenon. Our knowledge of primate intelligence is mainly based on laboratory studies of the problem-solving behavior of individual animals. Early experimental efforts to investigate social intelligence adopted the same experimental format. The upsurge of primate field studies in the nineteen sixties and the availability of stable captive groups in outdoor enclosures helped to change our perspective toward primate social intelligence, and underscored the need to consider more complex situations than those dealt with in earlier studies. Developing the empirical and conceptual tools to respond to this need is a major task for contemporary research on social intelligence.

INTRODUCTION

The concept of "intelligence" applied to the behavior of animals inevitably creates difficulties. Even when we are referring to a single species - namely, ourselves - expert opinion varies widely as to what the concept should encompass. Whatever the disagreements, however, no contemporary authority holds the view that human intelligence is a simple unitary trait.

Extending the concept of intelligence to include species other than man compounds the difficulties, and for fairly obvious reasons. One of the chief problems is the presence in different species of specialized abilities and skills that make untenable any simple ordering of species with respect to intelligence (4,6).

Nevertheless, "social intelligence" has been identified as an appropriate topic for discussion at this conference. The organizers evidently consider the concept of intelligence to have some utility, in spite of the ambiguities. I agree, at least to the extent that it refers to certain aspects of behavior that are of great biological interest and importance, and for which we have no better term.

To provide a background for this paper it will be helpful to review those attributes that I believe most people regard as the hallmarks of intelligence.

INTELLIGENCE FROM THE BIOLOGICAL POINT OF VIEW

Self-regulation is a fundamental property of life. An essential requirement facing any organism is to maintain its own integrity and equilibrium in the face of changing circumstances. Indeed, this requirement and the ability to fulfill it are defining characteristics of living systems. Behavior is a part of this process (16).

From a strictly functional standpoint, therefore, we may say that the two most fundamental themes in the evolution of behavior are knowing and wanting. All behaving organisms, from the most primitive to the most complex and highly evolved, act on the basis of information (knowing) and with reference to their needs (wanting). To accomplish this implies a third function in which information is evaluated with respect to need. These dimensions are formal properties of behaving organisms. They are logically implied by our conventional definitions of behavior as a biological phenomenon (5,18).

Thus, the mere fact that an individual is using knowledge to solve problems or achieve goals that relate to its vital needs does not characterize intelligence, nor make it an interesting biological problem. It is not the presence of wanting and knowing that challenges and intrigues us, but the nature of the processes subserving these functions.

What is the nature of these processes that characterize intelligence? In spite of the vague and pluralistic nature of the concept of intelligence, there appears to be considerable agreement as to its distinguishing features. These relate to what is known, the sources of knowledge, the motives and goals that influence behavior, and the means that are used to relate what is known to what is wanted (1,9,13).

1. What is known. Intelligence is characterized by the ability to respond differentially to a large domain of objects and events. The range of effective stimuli is broad, and responsiveness to differences among stimuli is well-developed. At the same time, as an apparent countermeasure against the problem of information overload, intelligence includes the ability to simplify diversity by selective perception, by perceptual constancies, by abstraction, generalization, and the formation of concepts, rules, and strategies.

2. Sources of knowledge. The larger the part played by individual experience in the acquisition and organization of knowledge, the more inclined we are to see evidence of intelligence. Intelligence is characterized by docility, the tendency to modify existing knowledge in response to changing circumstances. We are particularly impressed by evidence of "self-generated" knowledge, as reflected in innovative or creative solutions to problems. Integrating two or more disparate experiences or combining previously unrelated skills to produce a novel solution to a problem are hallmarks of intelligence.

3. Motives and goals. Intelligence is characterized by a diversity of motives and goals, and these often appear remote in form and function from procreation and the vital needs of the individual. The distinction between "appetitive" and "consummatory" actions, which is useful in describing the behavior of some organisms, is often difficult to make. Stimulus-seeking, curiousity, exploration, and play are prominent and assume the status of "independent" motives.

4. Means and ends. The means that are used in relating what is "known" to what is "wanted" are variable and often indirect. Motor patterns are flexible and highly differentiated, and they are expressed in a large repertoire of specific competences and skills. Evidence of planning, of foresight, of the ability to establish and work toward subgoals, to single out the essential features that define a problem, are the epitome of intelligence.

All of the foregoing attributes of intelligence are empirically established and readily demonstrable by objective means. They are more numerous, more prominent, and more consistently manifested in the behavior of some taxa than in the behavior of others. The degree to which they are present and the specific conditions under which they operate have an obvious and direct bearing on some of the central issues of biology: They affect the range of environmental conditions in which an individual can function (ecological niche), its capacity for dealing with changing circumstances (adaptibility), and its ability to survive, to reproduce, and to provide for its young (reproductive success). Thus, in spite of the ambiguities inherent in the concept of intelligence - indeed, the impossibility of ever defining it precisely - we can see that it serves the useful purpose of referring to certain emergent properties of evolution that are of fundamental importance.

PRIMATE INTELLECTUAL CAPACITIES

Our view of what constitutes intelligence is patently anthropocentric. However, it has also been heavily influenced by information on the nonhuman primates, most of which has been derived from laboratory research on individual animals. The dominant research tradition, which can be dated from the work of Köhler, Thorndike, and Yerkes at the turn of the century, is based on the assumption that intelligence can be revealed most effectively by confronting the individual with problems and recording its successes and failures as it attempts to solve them. The methods are straightforward, flexible, and extremely powerful. The approach has been enormously fruitful in generating ideas and solid findings. Our considerable knowledge of primate sensory and perceptual processes, of attentional factors, of the storage and retrieval of information, and of the ability to form concepts, learn rules, and develop problem-solving strategies and sets is the result of this approach. (For reviews see, e.g., (17,19)).

EARLY EXPERIMENTAL RESEARCH ON SOCIAL PROCESSES

Early systematic efforts to investigate social processes and social intelligence in nonhuman primates were carried out in essentially the same experimental format and under the same assumptions that had been so successful in the investigation of individual intellectual capacities. A problem or task was presented in a structured situation. The rewards or incentives for performing were usually nonsocial (e.g., food), and measures were obtained that reflected the individual's efficiency in achieving these goals. The major departure from research on individual intellectual capacity was that two animals were present and efficient performance required that the social element be taken into account.

A broad range of problems was explored: social facilitation, imitation, observational learning, food-sharing, and other forms of altruism, communication, and competition (including dominance testing). In some problems the partner's primary function was to provide information that was critical to a correct response (communication, imitation, observational learning); in other problems it could be considered an instrument (social "tool") that was essential to attaining the reward (cooperation, sharing, altruism); in still other problems it could be viewed as a barrier that had to be bypassed or surmounted in order to achieve success (competition, dominance testing). (For reviews see, e.g., (8,15).)

It should be noted that at the time most of this research was being done, information on the social behavior of primates living in natural groups was meager and unsystematic. The questions that were being asked were not inspired by observations of what primates actually did while living in social groups. As in the research on intellectual capacities, the emphasis was on theoretically interesting possibilities, on demonstrating potential achievements.

RECENT INFLUENCES

During the last fifteen years the character of thinking and research concerning primate social intelligence has changed dramatically. Many factors have contributed to this change. In the most general terms, however, I would ascribe the change to the ascendance of the naturalistic viewpoint. By that I mean a greater emphasis on phylogeny (including species-specific predilections and constraints), on behavioral ontogeny, on ecology, and a greater concern with behavior as an evolutionary product and as a contributing factor in the evolutionary process (9,11,12).

The early contributions of the ethologists were important. I have no doubt, however, that the most influential factor with respect to our views of primate social intelligence was field research. Field accounts revealed levels of social complexity that had scarcely been suspected previously. To cite just one example, Kummer's description of tripartite relations in hamadryas baboons not only drew attention to a high order of social intelligence operating in natural groups, but also pointed up the need to go beyond the exclusive focus on dyadic relations that had characterized the earlier experimental research (7). I believe it was also important that many behavioral scientists who had been trained in the experimental tradition had the opportunity to gain first-hand experience with free-ranging primate groups. For example, among the handful of experimental psychologists at the Yerkes Laboratories in Florida during the nineteen sixties, Berkson, Bernstein, Davenport, Menzel and I all became involved in field research.

Another important development during this same period complemented and reinforced the liberalizing influence of field research. For the first time in appreciable numbers, scientists had access to stable captive groups. In many cases these approximated the size and composition of natural groups; it was possible to investigate mating, the development of young, the importance of kinship, the formation of coalitions, and the dynamics of social change. Moreover, captive groups were often maintained in large enclosures, so that the intimate part played by spatial adjustments in the regulation of social life became apparent.

CONCLUSIONS

Research on the intellectual capacities of the nonhuman primates indicates that they are adept at integrating information in complex ways across space and time. There is good reason to suppose that these cognitive abilities play an important part in the regulation of social life, probably to a greater extent than for any other gregarious species.

Early attempts to demonstrate the ways in which high-level cognitive skills might influence social behavior yielded positive results. A broad range of problems was investigated, and the findings confirmed that the impressive cognitive abilities of monkeys and apes, initially revealed in nonsocial contexts, were indeed available and could be utilized for solving social "problems." At the same time, however, the fact that this research usually involved a single dyad working in a contrived, highly structured, task situation for nonsocial incentives (usually food) was a limitation that eventually had to be faced.

Among the more influential factors causing dissatisfaction with the constraints imposed by the early laboratory investigations of social cognition was the rapid increase in field studies, beginning in the nineteen sixties, and the changes in perspective that accompanied it. The field studies were a cogent reminder that most primates live throughout their lives in sizeable social groups, in the midst of many other individuals. Consequently, it was no longer possible in most species to treat the isolated dyad as a sufficient social "unit" in research on social processes and social cognition. The isolated dyad can be construed in terms of actor and reactor; events can be appropriately described in terms of "who-does-what-to-whom." The assumption of linear causality implied by this formulation becomes suspect when applied to dyadic interactions within the context of a social group. In a group setting, how one individual responds to another frequently depends on the other individuals present, on their arrangements in space, on their perceived moods, their relationships with the interactants, and so on. For example, a female's response to someone else's troublesome infant might very well be influenced by her relationship to the infant's mother, the mother's location, and the presence and location of the infant's older siblings. Note that in this hypothetical example, which is probably not at all atypical, it is not only the identity and location of other group members that influence the actor's response, but also her knowledge of their relationship to the infant and her assessment of their current moods and individual proclivities.

We need not confine ourselves to hypothetical examples: Young chimpanzees who have lived together in a large outdoor enclosure are able to infer from the behavior of two familiar leaders which of them is moving toward the larger or more preferred of two food sources (13). Rhesus monkeys living in a social group not only know their status vis á vis other members of the group, but also know the status of other group members in relation to each other, and can use this information to their advantage. For example: In a situation in which thirsty monkeys are competing for access to a drinking tube, subordinate monkey C will threaten dominant monkey B, while casting glances toward A (who is dominant over both B and C), with the effect that A starts to threaten B, causing B to abandon the tube, whereupon C approaches, and drinks. Thus, C uses his knowledge of the power relations between A and B to displace B, even though he is subordinate to B and would not threaten him if A were not also present (2).

As these examples suggest, one of the consequences of abiding social relationships is that the individuals become increasingly skillful in utilizing subtle cues indicating each other's intentions. This poses some difficult problems in behavioral recording methods that have yet to be resolved: Consider, for example, a mother monkey who has been for some days in the process of weaning her infant. She perceives the fact that it is facing toward her as the beginning of an approach, which she anticipates will culminate in an attempt to suckle. Her response to this perception is to turn her back upon the infant, which it perceives, in turn, as a rejection, whereupon it responds with distress vocalizations. Because each individual responded to its perception of what the other was likely to do, what might have been a lengthy, explicit, and highly-structured interchange is "short-circuited." An observer who was unfamiliar with the immediate history of the relationship and with its current "theme" would probably be hard-pressed to describe the effective stimulus that produced the infant's distress.

I trust these few examples are sufficient to demonstrate that complex cognitive abilities do indeed play an important and pervasive role in the regulation of social life in at least some primate species. It is reasonable to assume that individuals living within social groups spend much of their time monitoring their companions, gathering information about them, and elaborating social strategies and skills. These activities have direct and obvious consequences for the individual's competence as a social entity and carry clear implications for its ultimate biological success. In such mundane but vital matters as avoiding a fight or gaining access to food or a mate, they can make the difference between failure or success. That much is clear from current research (e.g., (2,8-11)). As yet, however, we lack a detailed and comprehensive understanding of the development and exercise of "social intelligence" for even one species.

In approaching this goal, I anticipate that laboratory research on primate social intelligence will remain firmly tied to the level of complexity manifested in naturally constituted social groups. These will provide the reference point, the source of the phenomena to be explained. Additional descriptive data will be required on social processes, particularly data that indicate the range, complexity, and subtlety of cognitive elements in social interaction. More refined normative descriptions will not be sufficient by themselves, however. It will also be necessary to experiment, to modify existing situations, to manipulate and control relevant variables. Experimental research will be performed within the context of established social groups, as well as in more restricted settings, as has been the case in the recent past.

A prerequisite for progress in both the descriptive and experimental sides of this endeavor is the availability of appropriate research tools. We will need a conceptualization of primate social processes that explicitly recognizes the importance of cognition, and a methodology expressly designed to deal with the cognitive aspects of social life.

With respect to conceptualization, I see a need to re-think the social environment from a radically different starting point than has been used in the past. The aim will be to establish new classes of independent and dependent variables, selected on the basis of their presumed relevance to social cognition. Thus, the social group will be construed in perceptual-field terms, in terms of classes of problems and types of opportunities, in terms of the content, quality, and consistency of the information available, in terms of the means the individual uses to gather and evaluate information and put it to use. More specifically, to what extent is an individual influenced by the spatial gestalt of the group, by the distribution of individual members and his knowledge of their moods and relationships? Is it useful to conceive of the group at any moment in time as a field of forces or vectors that have a determinate effect on the probable location and behavior of each individual within it? If this is not an appropriate way of viewing the group, what are some useful alternatives? Do individuals that convey little information, redundant information, or conflicting information occupy different positions in the attention structure of group members than those that are more likely to provide rich, new, or unambiguous information? How much and what kind of information does an individual gather from merely observing other group members interact with each other and with the nonsocial environment? Does he acquire information about ("knowing that") or specific skills ("knowing how")? Can we identify natural counterparts in a social setting of such higher-order cognitive achievements as discrimination learning-sets, oddity, and matching-from-sample?

Such a re-conceptualization requires that objective methods be available for recording events within a social group. In fact, some progress in this direction has been made (e.g., (2,13)). based on these beginnings, there is no reason to suppose that the requisite methods will be esoteric or difficult to use. The basic units for recording social behavior will be different from those now in common use, however, and will complement existing methods, rather than supersede them. To cite just one example from my own experience, it was apparent to me that the who-does-what-to-whom formulation, favored by most researchers, was missing an essential point about primate social behavior. This point was that social exchanges were episodic. All but the simplest interchanges have a beginning, a middle, and an end; they relate to some issue of importance to at least one of the participants, and they terminate with that issue either resolved to the satisfaction of one or both of them, or left hanging. In other words, the "unit" of social intercourse could be viewed as a transaction, which could be characterized in terms of its theme or content, its structure, its component processes, and its outcome. Denise Long developed a method to apply these ideas to mother-infant relations in two species of macaque monkeys, rhesus and bonnet. Her data have not been analyzed completely, but several findings are already clear: The species do not differ in total number of transactions or their developmental course. In rhesus, however, the proportion of conflictual transactions to total transactions is substantially higher than it is in bonnet monkeys. Further analysis of the data should reveal whether developmental and species differences exist in the techniques or strategies used by mother and infant to avoid or resolve conflictual transactions. One might even expect to show whether such strategies, acquired during the early socialization period, are used to deal with social conflicts in adulthood (2,10).

Acknowledgements. Preparation of this article was supported in part by grants HD06367 and RR00169 from the National Institutes of Health, USA. I am grateful for the comments received from other participants at the conference, particularly, J.H. Crook, D.J. Gillan, D.R. Griffin, M. Lindauer, G.H. Orians, and C.A. Ristau.

REFERENCES

(1) Anderson, B.F. 1975. Cognitive Psychology. New York: Academic Press.

(2) Anderson, C.O., and Mason, W.A. 1978. Competitive social strategies in groups of deprived and experienced monkeys. Dev. Psychobiol. 11: 289-299.

(3) Gottlieb, G. 1980. Recent history of comparative psychology and ethology. In Experimental Psychology at 100, ed. E. Hearst. Hillsdale, NJ: Erlbaum.

(4) Hodos, W., and Campbell, B.G. 1969. Scala naturae: Why there is no theory in comparative psychology. Psychol. Rev. 76: 337-350.

(5) Hull, C.L. 1943. Principles of Behavior. New York: Appleton-Century-Crofts.

(6) Jerison, H. 1973. Evolution of the Brain and Intelligence. New York: Academic Press.

(7) Kummer, H. 1967. Tripartite relations in hamadryas baboons. In Social Communication Among Primates, ed. S.A. Altmann, pp. 63-71. Chicago: University of Chicago Press.

(8) Mason, W.A. 1964. Sociability and social organization in monkeys and apes. In Advances in Experimental Psychology, ed. L. Berkowitz, vol. 1, pp. 277-305. New York: Academic Press.

(9) Mason, W.A. 1970. Early deprivation in biological perspective. In Education of the Infant and Young Child, ed. V.H. Denenberg, pp. 25-50. New York: Academic Press.

(10) Mason, W.A. 1978. Social experience and primate cognitive development. In The Development of Behavior: Comparative and Evolutionary Aspects, eds. G.M. Burghardt and M. Bekoff, pp. 233-251. New York: Garland Press.

(11) Mason, W.A. 1979. Ontogeny of social behavior. In Handbook of Behavioral Neurobiology: Social Behavior and Communication, eds. P. Marler and J.G. Vandenbergh, vol. 3, pp. 1-28. New York: Plenum Press.

(12) Mason, W.A. 1980. Minding our business. Am. Psychol. 35: 964-967.

(13) Menzel, E.W. 1974. A group of young chimpanzees in a one-acre field. In Behavior of Nonhuman Primates, eds. A.M. Schrier and F. Stollnitz, vol. 5, pp. 83-153. New York: Academic Press.

(14) Nissen, H.W. 1951. Phylogenetic comparison. In Handbook of Experimental Psychology, ed. S.S. Stevens, pp. 347-386. New York: John Wiley & Sons.

(15) Nissen, H.W. 1951. Social behavior in primates. In Comparative Psychology, third edition, ed. C.P. Stone, pp. 423-457. New York: Prentice-Hall.

(16) Piaget, J. 1971. Biology and Knowledge. Chicago: University of Chicago Press.

(17) Riopelle, A.J., and Hill, C.W. 1973. Complex processes. In Comparative Psychology: A Modern Survey, eds. D.A. Dewsbury and D.A. Rethlingshafer, pp. 510-548. New York: McGraw-Hill.

(18) Tolman, E.C. 1932. Purposive Behavior in Animals and Men. New York: Century Co.

(19) Warren, J.M. 1965. Primate learning in comparative perspective. In Behavior of Nonhuman Primates, ed. A.M. Schrier, H.F. Harlow, and F. Stollnitz, vol. 1, pp. 249-281. New York: Academic Press.

Animal Mind - Human Mind, ed. D.R. Griffin, pp. 145-158.
Dahlem Konferenzen 1982. Berlin, Heidelberg, New York: Springer-Verlag.

Internal Representation

L. A. Cooper
Dept. of Psychology and Learning Research and Development Center
University of Pittsburgh, Pittsburgh, PA 15260, USA

Abstract. In this paper, the issue of representation in cognitive psychology is considered. Emphasis is placed on the internal representation and processing of spatial information. A distinction between "analog" and "nonanalog" processes and representations is made, and a selection of empirical studies suggesting the need to postulate a special type of analog representational system is reviewed. The history of the debate over the necessity for postulating such a form of representation is outlined briefly. Finally, emerging directions in the study of internal representation are suggested.

INTRODUCTION

Issues of representation have been central in cognitive psychology for the past decade, and they form the core of the growing field of cognitive science. Researchers in areas as seemingly diverse as perception, pattern recognition, imagery, memory, categorization, and problem solving share the common desire to understand the nature of our mental models, knowledge, or internal representations of the external world. By "internal representation" we mean, quite generally, the content, structure, and organization of knowledge. Typical questions posed in the study of internal representation include: What properties of the world are incorporated in corresponding internal representations? How are these represented properties interrelated or organized? To what extent and in what sense are internal representations structurally or functionally isomorphic to their external counterparts? To what extent is external information recoded into an internal symbolic form, and how are such symbolic systems organized?

Central to an understanding of the nature of internal representation is a specification of the nature of the mental processes or operations that act upon, manipulate, and retrieve information from a representation. Indeed, Anderson (1) has recently argued that there is a fundamental indeterminacy in deciding issues of cognitive representation. That is, behavioral data alone cannot discriminate among proposed alternative knowledge representations, because one is simultaneously testing assumptions about the processes that operate on the proposed representation to produce the behavioral data. The best that one can do is to test a specified form of representation paired with a specified set of processes, or a representational system. While one might question the merit, force, and relevance of Anderson's argument, it does underscore the point that issues of internal representation and processing are closely intertwined.

Below, following a brief methodological note, I consider issues and research concerning the internal representation and processing of specifically spatial information. My treatment of area will be highly selective, owing to limitations of space. For other approaches to the problem of visual/spatial representation, see, for example, Palmer (13) and Posner (14). For relevant work in the area of artificial intelligence, see Winston (22), and for a developmental perspective see Liben, Patterson, and Newcombe (12). I have chosen this particular content domain for three reasons. First, I am more familiar with the work on spatial representation than work on the internal representation of linguistic information. Second, research on the representation of spatial information has been quite active recently, and the resulting evidence and controversies are relevant to more general issues of representation. Third, for purposes of a conference on animal mind and human mind a discussion of the representation of spatial information seems more appropriate than a discussion of the representation of verbal or linguistic information. The sorts of questions that investigators in the area of spatial representation are attempting to address might be exemplified by the following: How is it that an organism internally represents a familiar spatial environment for purposes of navigation? Is the internal representation like a "cognitive map" in which information about the distance between sets of landmarks is preserved? Or, is the representation more appropriately characterized as an abstract internal structure in which the relationships among landmarks are connected in the form of, say, an associative network? It seems equally appropriate for nonlinguistic animals and for humans to pose these possibilities as mental representations underlying locomotion.

METHODOLOGICAL NOTE

The difficulty of subjecting things as intangible as mental representations and operations to precise, objective study is notorious in behavioral science. Having no external access to internal representations and processes, we cannot directly detect even the occurrence of such events, let alone record and measure their properties. Instead, we must be content with the measurement of externally observable events and overt physical responses. In the case of the study of the representation and processing of spatial information, a very powerful technique - mental chronometry - has been developed over the past two decades and has been useful in making inferences about mental events from behavioral data.

The basic notion behind the use of chronometric techniques is that the time required to solve a spatial problem - when considered in conjunction with specified spatial properties of test stimuli - may provide information that places significant constraints on the nature of the intervening representations and processes used to solve the problem. In many of the studies to be reviewed below, the time needed to determine whether or not two visual test stimuli are identical has been shown to be systematically related to the extent of spatial displacement between the two stimuli. The claim that would be made on the basis of such a finding is that the underlying mental representation and processes must be of a class that would produce such an orderly relationship between time and spatial displacement. In addition to, and convergent with, chronometric methods, several investigators have recently begun to use other external indices of internal processing - most notably the pattern and duration of eye fixations on visual stimuli during solution of a spatial problem, e.g., Just and Carpenter (6) - in making inferences concerning the nature of underlying representations.

INTERNAL REPRESENTATION OF SPATIAL OBJECTS AND OPERATIONS

The principal theoretical issue that has been raised with respect to both the internal representation of spatial objects and the internal processing of such represented objects concerns the type and degree of correspondence or isomorphism between the represented objects and operations and their external referents. A central distinction that has been drawn is between "analog" representations and processes and "nonanalog" representations and processes. While there has been considerable confusion and disagreement concerning the precise meaning of these two terms, the following comments will probably not seriously misrepresent many investigators' theoretical positions: An analog representation or process is one in which the relational structure of external events is preserved in the relational structure of the corresponding internal representations. A nonanalog or symbolic representation or process is one for which this is not the case. To this definition of an analog representational process, Pylyshyn (18) has added another condition: A process is to be called analog only if the correspondence between the external and internal relational structures is a necessary consequence of intrinsic properties of the process. The relevance of this additional condition will become clear below.

A central issue in the ongoing debate between "analog" and "non-analog" theorists is whether it is necessary to postulate special types of processes and representations - namely, analog ones - for spatial information in addition to the discrete, symbolic representations which have been successful in characterizing knowledge of verbal/linguistic information. A number of lines of experimental work have suggested the need for such a special form of representation in the spatial domain, and two bodies of research are summarized below - work on "mental rotation" initiated primarily by Shepard, Cooper, and Metzler and work on the properties of mental images by Kosslyn and his collaborators. Reviews of these two bodies of research can be found in Cooper and Shepard (5), Shepard (19,20), and Kosslyn (7).

In a now-classic study, Shepard and Metzler (21) required human subjects to view pairs of perspective drawings of three-dimensional objects. On each trial, subjects had to determine whether the two objects were the same in shape or were mirror images. In addition to a possible difference in shape, the objects could also differ in orientation either in the two-dimensional picture plane or in depth. The central finding was that the time required to determine whether the two objects were the same in shape increased in a strikingly linear fashion with the angular difference between their portrayed orientations. Furthermore, the slope and the intercept of this linear function were virtually identical for pairs that differed by a rotation in depth and pairs that differed by a rotation in the picture plane. The linear increase in response time with angular difference between the objects led Shepard and Metzler (21) to suggest that the task was performed by "mentally rotating" an internal representation of one object into congruence with the other object and then comparing the two representations for a match or mismatch in shape.

This basic linear relationship between angular displacement, or extent of difference in orientation, and problem solution time is quite robust - holding up over a variety of stimulus modifications and task variations. For example, linear response-time functions have been found using alphanumeric characters and random two-dimensional polygons as stimuli. They have been obtained when a single rotated pattern must be compared with a pattern in memory, when rotations must be carried out before a test shape is displayed, and when the discrimination is changed to include subtly different distractors as well as mirror images. Special populations - most notably children and the blind (using tactual presentation) - also exhibit the linear relation between time and transformational distance. A review of these and other studies of mental rotation is provided in Cooper and Shepard (5).

Why have these studies demonstrating a linear relationship between time and spatial separation in the orientation of test stimuli been interpreted as suggesting analog representation of spatial objects and transformations? The reason is that the linearity suggests the correspondence between internal and external relational structures mentioned earlier as the chief characteristic of an analog process. That is, the linear relation between time and transformational distance suggests that the internal process used to solve this spatial problem passes through a trajectory or series of intermediate states, between the beginning and the end of the process, which have a one-to-one correspondence to the intermediate stages in the external rotation of an object. Furthermore, the linear response-time function indicates that the time needed to compare two visual objects at orientations A and C is an additive combination of the time needed to compare those objects at orientations A and B and the time needed to compare the objects at orientations B and C. This finding is indirect evidence supporting the claim that the internal process underlying comparison of the objects presented in orientations A and C passes through an intermediate state corresponding to orientation B.

We have argued (see, e.g., Cooper (4), Shepard (19) that this one-to-one correspondence between the intermediate stages of the internal process and the corresponding external process need entail only that the internal process passes through an ordered series of states at each one of which the individual is especially prepared for the presentation of the appropriate corresponding external object at the appropriate corresponding external orientation. This implies that a mental transformation of an object is an analog rotation if it can be demonstrated that the individual carrying out the transformation is internally representing the object in successively further rotated orientations in the sense that, during the process, that individual responds more rapidly and accurately to test stimuli in correspondingly more and more rotated orientations. Contrast this with a nonanalog internal transformation such as the matrix multiplication by which a computer might compute the new coordinates for a rotated system of points. This process would not be analog in that the intermediate stages of the calculation do not have a one-to-one correspondence to intermediate orientations in that the computer would not be in a state of readiness of the presentation of intermediate orientations at intermediate times.

A somewhat more direct experimental test of the analog nature of mental rotation has been provided by Cooper (4). Subjects were asked to imagine a shape rotating around a circle at a fixed rate. At some unpredictable moment during the rotation, a test shape was presented and subjects had to discriminate which of two versions of the shape had been displayed. Response times were shortest when the test shape was presented in the position that corresponded to the inferred orientation of the rotating internal representation. Furthermore, times increased linearly as the test shape departed by greater and greater angles from this "expected" orientation. These results are consistent with the idea that mental rotation is an analog of perceiving an actual rotation in that the internal and the external processes have the required relation of readiness for responding to the appropriate external object at intermediate stages.

Another line of evidence often cited in support of postulating analog representations and processes comes from the extensive work of Kosslyn and his collaborators. Many of Kosslyn's experiments are predicated on a logic similar to the studies of mental rotation discussed above in that the time required to make a spatial decision is related to extent of spatial displacement, and inferences concerning correspondences between intermediate stages of external and internal processing are made. (Kosslyn's experiments typically differ from the mental rotation studies in that subjects are explicitly instructed to form mental images as the basis for their decisions, whereas in the mental rotation experiments discriminative responses with no particular instructions are required.)

Consider, for example, studies on the scanning of mental images (see, in particular, Kosslyn, Ball, and Reiser (9)). In these experiments, subjects are asked to generate visual images of objects or of spatial layouts, and they are asked to focus attention on one particular location in the object or layout. They are then required to verify, as rapidly as possible, whether another property or object is present in the object or layout. What Kosslyn finds - under these particular instructional conditions - is that the time to make the verification is an increasing function of the distance between the attended property or location and the property or location to be verified. The parallel between these findings and those of the mental rotation studies, in which transformational distance of orientation difference was the chief independent variable, is apparent, and a similar logic concerning the underlying representations and processing mechanisms can be applied. Furthermore, these results were obtained even when potential confoundings - such as the number of locations or properties intervening between the attended and the probed locations - are controlled. These results, then provide additional evidence that an internal representation of space - when subjected to the transformation of scanning - has a structural correspondence to the external depiction of distance between objects or properties which is required of an analog representation. (The results of Kosslyn's research program go far beyond these image scanning findings, but the findings are too extensive to detail in this paper.)

One of the chief contributions of Kosslyn's program of work has been the development of a theory of image processing and representation that is embodied in a running computer simulation model (see Kosslyn (7, 8) and Kosslyn & Shwartz (11)). The details of the simulation model are too numerous to discuss here, but suffice it to say that the model combines a quasi-pictorial "surface representation" to which various operations such as scanning and rotation can be applied with an underlying symbolic "deep representation" in long-term memory from which images depicted in the surface representation can be generated. In retaining the concept of a surface representation for which there is a correspondence between the relative spatial characteristics (e.g., interobject or interproperty distances) of external objects and internal representations of those objects, Kosslyn's model clearly argues for the necessity of postulating analog representations and processes, in the sense described above.

Despite the experimental evidence from the mental rotation studies and the results of Kosslyn's research program, the view that an analog representational system is needed to characterize spatial representation and transformation has not gone unchallenged. The precise nature and focus of the debate between analog theorists and those who argue that a single symbolic representational system is adequate for describing all aspects of human cognition has changed over the past decade, sometimes in rather subtle ways. Only the central points of this exchange will be outlined briefly here.

Early criticisms of the use of analog representations as explanatory constructs (in particular, Pylyshyn (15) pointed to the incoherence of the "picture metaphor" for mental imagery and to the necessity for postulating a single, underlying symbolic representational system for translating between spatial and linguistic concepts. Another line of attack has consisted of showing that symbolic representations and processes - however unnatural and ad hoc in nature - can be made to imitate the phenomena generally used to support the need for analog representations and operations (see Anderson (1) and for a different point of view, Kosslyn and Pomerantz (10)).

More recently, Pylyshyn (16-18) has voiced a rather different set of objections to the research and theoretical efforts of proponents of the analog position. In addition to taking issue with various features of and claims for the Kosslyn and Shwartz (11) model, Pylyshyn has argued that the results of mental transformation experiments may be accounted for by subjects' use of tacit knowledge concerning the nature of physical transformations and their interpretation of experimental situations as inviting the simulation of what they know the appropriate physical transformation to be. This is in contrast to the view that so-called analog operations are knowledge-independent properties of an internal mechanism that must necessarily be used to solve rotation and scanning problems. Pylyshyn argues further for this "tacit knowledge" account by demonstrating that results of mental rotation and scanning experiments can be influenced by factors such as instructions, belief, and practice, and hence, are "cognitively penetrable." The final step in his argument is to hold that cognitively penetrable processes must make reference to underlying symbolic operations in that knowledge is shown to influence experimental outcomes.

(Kosslyn (8) has recently provided a rejoinder to Pylyshyn's criticisms of his theory and to the arguments that an appeal to tacit knowledge offers an adequate account of all of the phenomena ascribed to analog representations and processes. His criticisms of the tacit knowledge position in essence are the following: First, results consistent with an imagery or analog representation account can be obtained when subjects are not explicitly told to generate images or to engage in processes analogous to external spatial transformations. Second, subjects will often exhibit performance that indicates that their internal representations of spatial objects have properties similar to those found in perceptual experiments which are counterintuitive and thus, can probably not be ascribed to tacit knowledge about physical objects and events.)

To sum up briefly the status of the debate concerning the need for postulating a special analog representational system for certain spatial information and transformations, the jury is still out. Furthermore, it is unlikely that a clear verdict will be reached in the near future. Nonetheless, this ongoing controversy has shifted emphasis over the past several years, indicating that some progress and convergence in thinking has been achieved. It has also effected tangible positive outcomes - among them the development of a clear, precise model of analog representation and processing by one theorist (Kosslyn) and the acquisition of a rich set of empirical findings to serve as a foundation for theory and further prediction.

EMERGING DIRECTIONS IN THE STUDY OF SPATIAL REPRESENTATION AND PROCESSING

I conclude this paper with a brief indication of two issues in the internal representation of spatial information that are currently receiving attention and which seem to constitute avenues for future research. The first of these issues, alluded to at the beginning of the paper, concerns the internal representation of large-scale spatial environments. (Note that most of the research reviewed in the body of this paper involves the representation and transformation of individual spatial objects, rather than the representation of the relationship among various spatial objects, locations, or events.) In this emerging field of study - termed "cognitive mapping," or "locational cognition" by various investigators, e.g., Chase and Chi (2) - the sorts of questions that are being posed include: Do representations of the environment preserve metric information much like a map would? Is information about the environment organized in a hierarchical fashion? How is information about the environment accessed in order to plan and execute a route from one location to another? A second issue concerns changes in the representation of spatial information with learning. One particular focus of this research contrasts the spatial representations of "experts" with those of "novices." For example, Chase, Chi, and Eastman (3) are currently conducting an extensive study comparing the spatial representational structures and processes of expert taxi cab drivers with those of less experienced persons. Their results to date suggest that experts and novices differ in their modes of accessing information about alternative spatial routes in well-defined ways.

One exciting prospect of this general line of research - applied to not only spatial representation, but other forms of representation as well - is the possibility of extracting general principles concerning changes in the organization and nature of knowledge structures with experience.

REFERENCES

(1) Anderson, J.R. 1978. Arguments concerning representations for mental imagery. Psychol. Rev. 85: 249-227.

(2) Chase, W.G., and Chi, M.T.H. 1981. Cognitive skill: Implications for spatial skill in large-scale environments. In Cognition, Social Behavior, and the Environment. Potomac, MD: Erlbaum, in press.

(3) Chase, W.G.; Chi, M.T.H.; and Eastman, R. 1980. Spatial representations of taxi drivers. Paper presented at the meeting of the Psychonomic Society, St. Louis.

(4) Cooper, L.A. 1976. Demonstration of a mental analog of an external rotation. Percep. Psychophys. 19: 296-304.

(5) Cooper, L.A., and Shepard, R.N. 1978. Transformations on representations of objects in space. In Handbook of Perception: Perceptual Coding, eds. E.C. Carterette and M.P. Friedman, vol. 8. New York: Academic Press.

(6) Just, M.A., and Carpenter, P.A. 1976. Eye fixations and cognitive processes. Cog. Psychol. 8: 441-480.

(7) Kosslyn, S.M. 1980. Image and Mind. Cambridge, MA: Harvard University Press.

(8) Kosslyn, S.M. 1981. The medium and the message in mental imagery: A theory. Psychol. Rev. 88: 46-66.

(9) Kosslyn, S.M.; Ball, T.M.; and Reiser, B.J. 1978. Visual images preserve metric spatial information: Evidence from studies of image spanning. J. Exp. Psychol.: Gen. 4: 47-60.

(10) Kosslyn, S.M., and Pomerantz, J.R. 1977. Imagery, propositions, and the form of internal representations. Cog. Psychol. 9: 52-76.

(11) Kosslyn, S.M., and Shwartz, S.P. 1977. A simulation of visual imagery. Cog. Sci. 1: 265-295.

(12) Liben, L.S.; Patterson, A.H.; and Newcombe, N. 1981. Spatial Representation and Behavior Across the Life Span. Theory and Applications. New York: Academic Press.

(13) Palmer, S.E. 1978. Fundamental aspects of cognitive representation. In Cognition and Categorization, eds. E. Rosch and B.B. Lloyd. Hillsdale, NJ: Erlbaum.

(14) Posner, M.I. 1978. Chronometric Explorations of Mind. Hillsdale, NJ: Erlbaum.

(15) Pylyshyn, Z.W. 1973. What the mind's eye tells the mind's brain: A critique of mental imagery. Psychol. Bull. 80: 1-24.

(16) Pylyshyn, Z.W. 1979. Imagery theory: Not mysterious - just wrong. Behav. Brain. Sci. 2: 516-563.

(17) Pylyshyn, Z.W. 1979. The rate of "mental rotation" of images: A test of a holistic analogue hypothesis. Mem. Cog. 7: 19-28.

(18) Pylyshyn, Z.W. 1981. The imagery debate: Analogue media versus tacit knowledge. Psychol. Rev. 87: 16-45.

(19) Shepard, R.N. 1975. Form, formation, and transformation of internal representations. In Information and Processing and Cognition: The Loyola Symposium, ed. R.L. Solso. Hillsdale, NJ: Erlbaum.

(20) Shepard, R.N. 1978. The mental image. Am. Psychol. 33: 125-137.

(21) Shepard, R.N., and Metzler, J. 1971. Mental rotation of three-dimensional objects. Science 171: 701-703.

(22) Winston, P.H. 1975. The Psychology of Computer Visions. New York: McGraw-Hill.

Animal Mind - Human Mind, ed. D.R. Griffin, pp. 159-176.
Dahlem Konferenzen 1982. Berlin, Heidelberg, New York: Springer-Verlag.

Problem Solving

G. Lüer
Institut für Psychologie der Rheinisch-Westfälischen Technischen Hochschule, 5100 Aachen, F. R. Germany

Abstract. The empirical investigation on thought process started at the beginning of our century. The research on problem-solving processes originated from approaches of Gestalt psychology and continues at the present time under the aspect of information processing.

Recognizing that different problems require the application of different problem-solving processes in man, classifications have been proposed describing different problem structures.

Information-processing procedures may be described on two levels: the level of knowledge and the level of heuristic methods. Knowledge as well as heuristic methods are kept and stored in different memories having different characteristics. It is assumed that the structure of the memory can be described by means of semantic networks.

To date empirical investigations of problem-solving processes have mainly been applied to less complex problem structures. Only recently have problem-solving processes been examined in highly complex reality domains.

Further topics of empirical research deal with changes of problem-solving processes evoked by training methods and with investigations on correlations between the performance on intelligence tests and problem-solving ability. Still others deal with information-processing procedures during analogical reasoning.

Problems of methodology are often associated with the process of data collection and with the analysis of these process data.

INTRODUCTION

Research on human thinking has been conducted from different viewpoints. Considering the research periods on human mental processes, the one on problem solving was significantly successful. The roots of this research tradition go back to the so-called "Würzburger Schule" under the leadership of Külpe and later German Gestalt psychology. It was given the name "productive thinking."

Today, we seem to be beginning a new phase of research on cognitive processes. This research is no longer limited to problem solving itself. Rather the aim is to include the influences from emotional and motivational states in addition to the more elementary processes of cognitive activity, such as recognition or structures of the memory. This is called "cognitive psychology" (10), or in a wider sense, "cognitive science" (13).

The main goal of this paper will be to summarize procedures which are presently used in psychology to investigate, describe, and explain principles of the human mind. Results from empirical investigations will complete the picture of present research. It is still questionable whether these principles and results are applicable to the animal mind.

BRIEF HISTORICAL REVIEW

The change toward a more systematic investigation of thinking processes took place at the beginning of our century and was due to two factors: (a) discontent over the extremely simplified explanations of mental activities given by the association theory, originally introduced by Aristotle (this theory describes elements of the mind (ideas) and their relationships to each other by associations), and (b) a variety of empirical data concerning the process of thinking which could no longer be explained by the association theory.

The so-called "Würzburger Schule" thoroughly analyzed records of self-monitoring from persons trying to solve problems. On this empirical basis the school ascertained that the course of thinking does not run erratically. These investigators recognized the effect of "determining tendencies" which were supposed to explain a directed activity satisfying the cognitive requirements. In addition, the idea that conscious images were the only basic elements of mental processes had to be abandoned.

The first period of psychological research on human thinking culminated in the essays of Selz (16). He strongly objected to the psychological interpretation of cognitive processes on the basis of associations. His approach to productive thinking included consideration of the following factors, the significance of which is quite obvious to us today: the importance of the problem, the importance of the problem structure, and the distinction of different mental processes in terms of productive thinking.

Gestalt Psychology studies became another important basis for the modern psychology of problem solving. Duncker's (3) analyses of protocols, which were obtained from subjects thinking aloud, resulted in subtle description of thinking processes. His considerations led to the understanding of thinking as a transformation process performed by the individual, and he described possible properties of problem states and defined the problem-solving process as the search for ways to change these properties. Impulses for the cognitive activities are released by an obstacle which is experienced by the individual as a barrier prohibiting an immediate change of the problem state. As a means to overcome this barrier, the problem solver uses so-called heuristic techniques. Duncker describes them, e.g., as conflict analysis, situational analysis, material analysis, and goal analysis. His description greatly influenced the modern way of understanding problem solving as information processing within the individual. Subjects confronted with problems use these heuristic techniques in order to analyze the given problem and to find solutions.

A listing of all theoretically possible ways to solve a problem is described by the picture of a solution tree. The different problem states to be transformed are represented in terms of branches of a tree.

The research on problem solving in respect to information processing reached its height with the application of computers. These could be used as effective systems of problem solving by manipulating symbols and could be programmed as models for mental processes. The basic idea of this new line of research was the assumption that men would function as information-processing systems when solving problems. According to Newell and Simon (11), such a system includes the elements as depicted in Fig. 1.

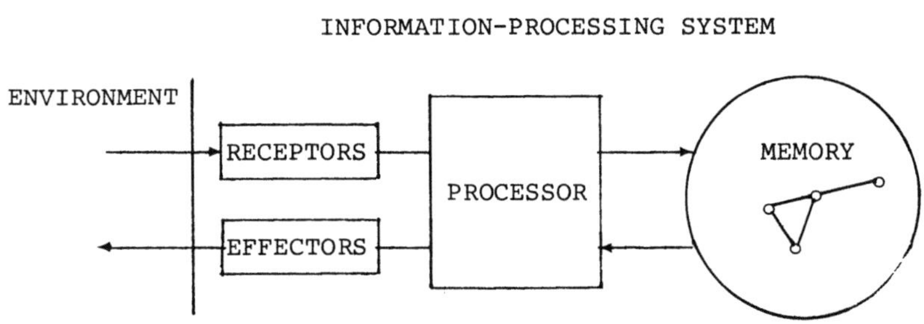

FIG. 1 - An information-processing system (from (11)).

Important components of problem solving are outlined in this diagram: the environment as the reality domain for problems, the processes of recognizing the attributes and possibilities for decoding and encoding them, the processes of manipulating symbols as the changes of the properties of problems, and the procedures for storing symbolic representations.

The composition of cognitive processes is more complex, since many different psychological processes have to be considered. Norman (13) proposed a list of aspects which are substantial for mental activities and reactions. Such a list includes known processes which have to be assumed to guide cognitive processes. The linkages and interactions between the elements are almost completely unknown. Figure 2 represents only a preliminary map for cognitive activities (13), explicitly showing which psychological processes have to be taken into account when

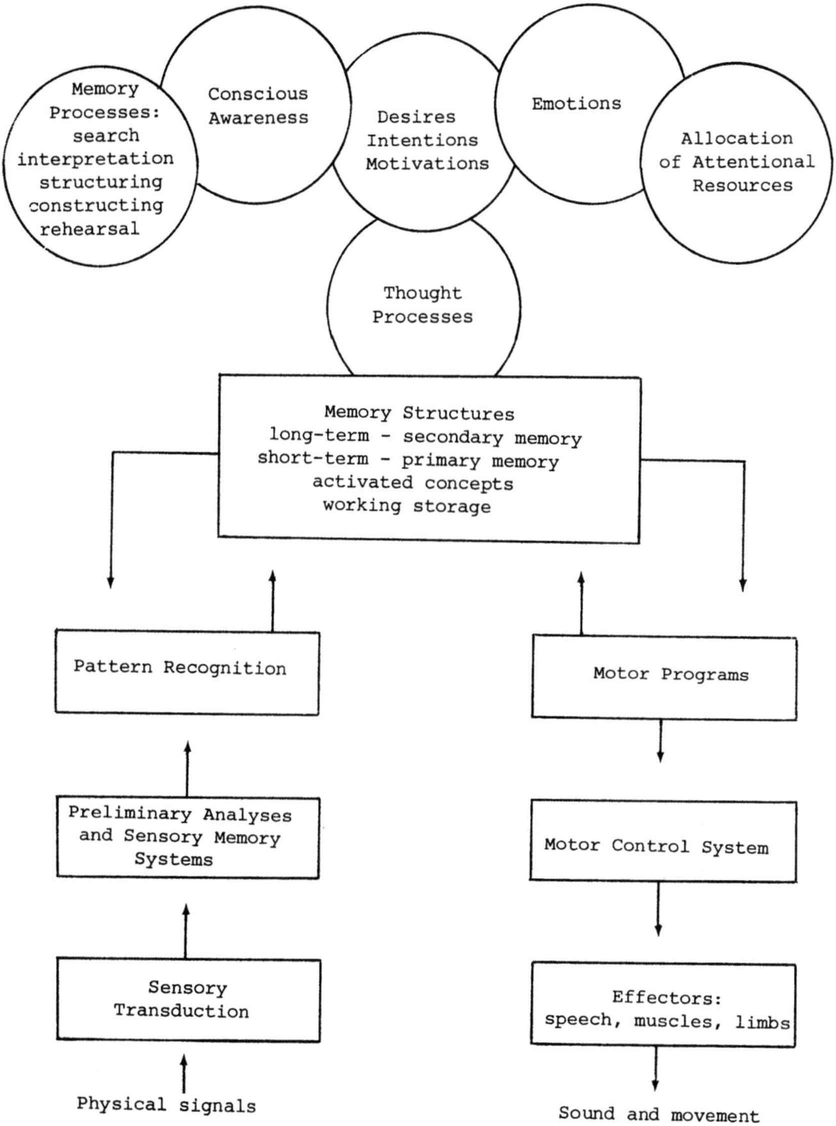

FIG. 2 - Flow chart of the human information-processing system (from (13).

studying cognition. These processes could be helpful in developing strategies for research on the animal mind.

DESCRIPTION OF PROBLEMS

Today we generally assume that different types of problems require different cognitive activities in the human mind. The classification of problems is thus an important part of psychological research. It represents a taxonomy of the requirements a problem solver has to meet. For this reason studies on problem-solving abilities of a certain species should begin with an analysis of the requirements the animals have to meet.

In the literature, there is a general consensus about the following components that characterize a problem situation: an unsatisfying initial state, a desired final state, and an obstacle (called a barrier) that prevents the transformation from the initial state into the final state. In spite of such common properties of problems, there are still great differences between various kinds of problems. The differences not only refer to the content of problem situations but also, e.g., to the clearness by which the initial and final states can be defined. Moreover, a problem is characterized by the type of barrier which has to be overcome.

Klix (6) and his co-workers (8) have examined subjects who were trying to solve problems. They described the activities of their subjects during problem-solving experiments in terms of heuristics. For that purpose they mainly used the game "Tower of Hanoi." All possible positions that could occur during this game can be depicted in a graph which corresponds to a complete solution tree as suggested by Duncker (3). In this case, problem solving of human subjects was studied on the background of an objective problem structure.

There are only a few problems that can be completely depicted in a formally exact way. Thus it is necessary to have available other classifications for problems. Attempts have been made to find psychological classifications for problems which cannot be formally depicted. These psychological classifications only make use of certain typical aspects of the problems.

The most comprehensive attempt to classify problems in a psychological manner was made by Dörner (1), whose contributions include typologies for problems which may be described according to the barriers to be overcome. His classification scheme for problems includes the dimensions "clearness of the goal criteria" and "degree of familiarity of means," yielding a fourfold table with different types of barriers (Fig. 3).

		clearness of the goal criteria	
		high	low
degree of familiarity of means	high	interpolation barrier	dialectic barrier
	low	synthesis barrier	dialectic and synthesis barriers

FIG. 3 - Classification of types of barriers for problems (from (1)).

Problems arise within a certain reality domain. In this sense the environment is a reality domain from which problems can arise, e.g., the need to improve an ecological situation. By applying operators - these being part of a reality domain just as are the state of affairs - new states of affairs are created.

State of affairs of a problem can be described and distinguished by dimensions. Examples for dimensions are: complexity, dynamics, transparency, degree of cross-linking, and existence of free components.

Operators can be attached to a certain reality domain. Dörner (1) proposes such classificatory dimensions as effective range, reversibility, complexity, and costs of application.

From this short discussion of problem classification, it should be clear that the various requirements imposed by different types of problems can only be dealt with by assuming a complex system of information processing in the human mind. Indeed, a highly flexible system that will meet all the various requirements should be necessary. That is why theoretical assumptions about the cognitive organization in human problem solvers have been established. These allow us to describe the possible cognitive structure of the human mind. Referring to Gillan's paper (this volume), one has to suggest similar forms of organization of cognitive abilities in apes.

HYPOTHESES ABOUT THE COGNITIVE STRUCTURE

The Two-Level-Hypothesis

By surveying the literature, Kluwe (7) found it to be widely assumed that the cognitive structure is divided into two parts: a structure of knowledge and a structure of heuristic methods. This hypothesis states that the information-processing procedure in the human mind takes place on two distinguishable levels.

In the structure of knowledge, data about states of affairs of reality domains are stored in memory, together with operations and their corresponding operators. This internal structure represents ideas about states of affairs and ways to change them. The ability to learn various forms of new behavior clearly demonstrates the existence of a structure of knowledge in animals.

The second processing level that is highly important for the problem-solving process is called heuristic structure. It includes procedures to solve problems which can be applied on the basis of the data from the knowledge level. Duncker's heuristic techniques are examples of mechanisms making up this structure. It seems to me questionable whether animals have such comparable techniques at their disposal.

Thus, problem solving is understood as a transformation process based on knowledge, with heuristic strategies serving as procedures for constructing solutions. The heuristic techniques form the program for operators which change the problem states to achieve the solution. The two-level-hypothesis describes just this interaction of knowledge about states of affairs and about operators to change them on one hand, and solution programs on the other. This hypothesis suggests a potential model for the structure of the information-processing system in the human mind.

Importance of Memory for Problem Solving

All hypotheses about a structured cognitive apparatus assume the existence of some sort of storage elements which explain the ability to memorize. Processes of problem solving are understood as information processing procedures operating on structures of memory. The following processes are of importance: (a) the representation of the problem in memory, (b) the changing of problem representation by learning processes, and (c) representation of simple and complex (heuristic) operations in memory.

It has been shown that problem-solving processes have to be considered in relation to different memory performances. This has led to the assumption that memory involves distinct steps. A frequently used classification assumes three categories: (a) the sensory memory (SM), (b) the short-term memory or primary memory (STM), and (c) the long-term memory or secondary memory (LTM).

The SM has quite a large capacity, but information can only be retained for a short period of time (about 100 to 400 ms). In contrast, the STM has only a small storage capacity, but allows for the acceptance of information at a high speed. The capacity is about 2 to 3 bits. The information can be retained for only 30 to 40 seconds. It is still controversial whether the information is then deleted by a process of self-decay or by interference with new information.

The LTM seems to have a large storage capacity but a low speed of information storage. In this part of memory, the actual knowledge about states of affairs and about ways to act are stored. Its content is fairly well-protected against loss of knowledge (forgetting). According to our present understanding, knowledge in the LTM is strongly linked to semantic units. From this idea one concludes that it would be possible to discover the organization principles of the LTM by means of semantic analyses.

One of the important classifications for the examination of problem-solving processes is systematology, as proposed by Krause and Krause (8). This includes: (a) a working memory (WM), as the active part of the LTM where problem solving mainly takes place; (b) a semantic memory (SEM), as the part of the LTM where the knowledge of the problem solver is represented (structure of knowledge); and (c) an operative memory (OM), as the part of the LTM where operators and heuristic strategies are stored (processing structure).

Investigating the structure of the memories is as important as distinguishing the different types of storage. The fact that information stored in memory can be quickly retrieved has led to the assumption of active networks in which units of knowledge in the form of nodes are linked to each other. Filing of information is done in the form of concepts. Their linkages are relations such as part-whole relations, subordinate-superordinate relations, and others. According to this model, nodes can be linked to each other by new relations, leading to an expansion of knowledge. A problem solver attempts to solve a still unsolved problem on the basis of content and relations stored in the network.

To illustrate the relation between the knowledge and the heuristic structure, Dörner (1) uses the so-called cuttle-fish hypothesis. This fish tries to catch knots of a fisherman's net by means of its eight tentacles and, starting at three knots, for example, it pursues superimposed and subordinate relations. If it finds a knot where these three threads come together, it has found a superimposed concept. Otherwise, it has to take needle and thread in order to tie new relations.

EMPIRICAL INVESTIGATIONS OF PROBLEM-SOLVING PROCESSES

The experiments in the field of artificial-intelligence research, which attempt to model human problem-solving processes on computers, began with the work of Newell and Simon (11). This so-called General Problem Solver (GPS) represents an information-processing system which is able to produce solutions to various problems on the basis of a given amount of information. The authors have further claimed to have simulated human problem solving using computers in a general sense and to have developed a system of information processing, which could serve as model for human information processing. It was found, however, that the simulation of single protocols of human problem solving is extremely tedious. In addition, it had to be recognized that the correspondence between simulation models and actual human information processing could not be achieved to a satisfying degree.

More or less experimental investigations with simple or at least completely transparent problems led to numerous new empirical results. Examples of some of these investigations are the analyses of problem solving in connection with propositional calculus problems (9), experimental investigations by means of the Tower of Hanoi, and cryptarithmetic tasks (6,11). These studies gave answers to questions of the following kind: Which elementary information-processing units constitute problem-solving strategies? Flow charts for elementary cognitive processes were constructed. How do organization forms of mental processes develop? How are problem-solving strategies generated and how are they used? To what extent do problem solvers deviate from logical thinking, and what are the conditions for such forms of illegal thinking?

During recent years important extensions of the idea of a general problem solver have been developed. Simon (17) made up a list of five new directions which were developed in the area of modeling the human mind. The main aspects of these directions are: (a) Different forms of attention to perceptual clues lead to different solution strategies. (b) The knowledge of experts has been studied. These results influenced the construction of new simulation models. (c) Models have been constructed in other semantically rich domains (e.g., in domains within physics). (d) New formalisms, called "production systems," explained parts of cognitive activities in a new light. (e) Processes underlying the understanding of natural language instructions have been modeled.

In addition to these findings about elementary information-processing processes, it was observed that the empirical results depend strongly on the types of problems under investigation. Thus, different types of problems are solved by means of different processing processes and different principles.

From all this clearly fruitful research on problem solving, the following conclusions can be drawn: At present, we know a great deal about the basic information-processing procedures of the human mind. However, we have to recognize the high degree of variability in their organization. This makes it quite difficult to formulate a _general_ theory of problem solving.

Besides this pure expansion of our knowledge, the empirical research on problem solving necessarily involved a variety of methodological problems, which are especially important with respect to the methods of data collection. Even behaviorally-oriented psychologists have complained that only rarely are mental processes directly coupled to observable behavior.

What data are available from problem-solving experiments? I will mention only the following three examples: data from verbal protocols, measurement of time requirement, and registration of eye movements. Since the days of the "Würzburger

Schule," psychologists in this field have had to face criticism because of the methods they use, the reason being that these methods are designed to provide information about the originating process that leads to the solution of a problem. The resulting data include protocols obtained from the problem solver thinking aloud, as well as the records of all the intermediate results and of observation data about the usage of applied tools. "Pure" experimenters may never be able to excuse cognitive psychologists for using such "soft" data as verbal records. Nevertheless, the information from these data sources proved to be relatively important. In addition, new contributions dealing with methodological problems of protocol analysis were published (4,13). Of course, it should be mentioned that protocols obtained from subjects thinking aloud are incomplete and can only be used to a limited extent, and with considerable effort, for a quantitative analysis.

Investigations about the time required for strictly defined information-processing techniques were carried out with a high degree of experimental precision. Sternberg (18) proposes a "component theory" for analogical reasoning with the following components of information processing: encoding, inference, mapping, application, and response. In addition, rules about the linkages of these processes into strategies were established, and efforts were made to generalize the "component theory" so that it could be applied to a variety of contents and formats. The resulting picture of analogical reasoning is theoretically sound and empirically well grounded. However, these data were collected from a highly selected sampling of subjects who first had to go through intensive training in order to be highly skilled in solving "analogy"-tasks.

Investigations of the correspondence between eye-movements and problem solving were carried out with a precision similar to that achieved in the experiments (18) mentioned above. The following data can be registered: saccades (gross eye-movements), loci, and times of fixation. As we can see from Putz-Osterloh's (14) careful analyses of eye-movements while subjects were engaged in the reasoning-items in intelligence tests, strategies can be derived which play a role during the information gathering.

Unsolved methodological problems in the research of problem solving include the following:
(a) Protocols of verbal statements cannot be abandoned, at least not for complex problems even when different types of data are registered during the problem-solving process. Protocols yield important information about the course of the cognitive process. What remains still unsolved is the problem of interindividual differences in the ability and willingness to report conscious mental processes to the experimenter.
(b) Investigations of long-term problem solving, e.g., with complex problems, generate a large amount of data which then cause extreme difficulties in data analysis. This is especially true for verbal protocols.
(c) Sufficient statistical methods for analyzing the course of the processes are still lacking.

At the end of my paper I will briefly summarize topics of present research in the field of human problem solving. During the last few years intensified activities in three new fields of the research of problem solving have been observed: (a) expansion into problems encountered in highly complex reality domains, (b) changes in cognitive processes by means of training in problem solving, (c) correlation between performance in an intelligence test and success in problem solving.

Dörner and co-workers (2) have extended the research on problem solving to very complex and dynamically developing reality domains. They allow subjects to control a system which simulated the civil administration of a middle-sized town. The system was established on a computer and is characterized by a high degree of cross-linking between the internal system variables. Important results of these experiments are: descriptions of the course of long-term problem-solving processes and of operations used by the subjects in order to solve particular problems, and the effect of the problem solver's motivation on the problem-solving process.

Training methods in problem solving which go beyond simple practicing of familiar procedures provide an opportunity to use knowledge about mental operations. With these training methods one can attempt to induce more effective problem solving in subjects. From these very complicated, and therefore not very numerous, investigations, two main points can be derived: (a) problem solving can indeed be improved, and (b) those modification methods which bring about the development of the heuristic structure and therefore lead to the formation of new heuristic techniques are particular effective.

Under experimental conditions one training method has proven to be effective. It is called the self-reflection method (5) and deals with promoting the development of problem-solving strategies. During the training period the problem solver is instructed to repeatedly ask questions concerning his own process of problem solving. These questions refer to his own position within the solving process, to the tools he has used so far, and to the difficulties he has encountered. The effectiveness of these training methods could be tested experimentally with respect to successful problem solving and to the organization of information-processing techniques.

It is difficult to understand why research on intelligence and problem solving have developed separately for such a long time without influencing each other. Only recently have empirical investigations demonstrated that performance in an intelligence test and success in problem-solving situations are not correlated. This logically absurd result motivated cognitive psychologists to make intensive efforts in the field of intelligence measurement (15). As a first step, problem requirements as they occur in intelligence tests were investigated. The following research on the solving processes which are required in intelligence tests shows what discriminates the successful from the unsuccessful problem solver. In a second step, the missing correlations between the two cognitive performances were investigated. By varying the characteristics of problems - such as transparency of the variables in problems and semantic versus nonsemantic conditions of the task - it could be demonstrated that the items found in intelligence tests are much simpler than those occurring in complex problems, as far as the requirements to the problem solver are concerned. When complex problems are simplified in such a way that additional information about their structure is given in a clear and vivid manner, correlations between the performance in problem solving and in intelligence tests can be systematically demonstrated. Obviously the tasks of an intelligence test represent only a subset of problem requirements. In any case, these requirements still lag far behind those of complex and dynamically developing problems. This would lead to the conclusion, however, that so far, in measuring intelligence, one has apparently dealt only with problems of a simple nature.

OUTLOOK

The research on human problem-solving processes is one of the fields in psychology where many new insights have been gained during the last two decades. Progress has been made in empirical research as well as in theoretical explanations of cognitive processes. This promotes our understanding of information-processing processes underlying problem-solving abilities.

With the incorporation of complex and dynamic problem situations into research, it was recognized that the former approaches had often been somewhat limited and biased. Up to now, research into problem solving has mainly been carried out from mentalistic and rationalistic viewpoints, reminiscent of earlier periods of philosophy. Thus, we have a situation in cognitive psychology where the motivating and emotional states of the problem solver are hardly taken into account. This situation is really not understandable, especially as every psychologist doing empirical cognitive research has often observed fluctuations in emotional and motivational states of his subjects.

They appear as oscillating states of the problem solver with regard to his interest in the task, his ability and willingness to concentrate, his enjoyment or dislike of searching for a solution, and his level of exertion vis-a-vis the given requirements. Norman (13), in his "twelve issues," mentioned some additional psychological processes which influence the process of information processing. It is much to be hoped, that more will be done than just the listing of them.

Acknowledgements. This work was supported by the Deutsche Forschungsgemeinschaft, Bonn-Bad Godesberg, F.R. Germany. The author is grateful to E. Eisermann for translating the text and to G. Trommsdorff for valuable comments on the manuscript.

REFERENCES

(1) Dörner, D. 1976. Problemlösen als Informationsverarbeitung. Stuttgart: Kohlhammer.

(2) Dörner, D. 1980. On the difficulties people have in dealing with complexity. Simul. Games 11: 87-106.

(3) Duncker, K. 1935. Zur Psychologie des produktiven Denkens. Berlin: J. Springer.

(4) Ericsson, K.A., and Simon, H.A. 1980. Verbal reports as data. Psychol. Rev. 87: 215-251.

(5) Hesse, F.W. 1979. Alternative Ansätze zur Entwicklung heuristischer Strategien für den Bereich des schlußfolgernden Denkens. In Komplexe menschliche Informationsverarbeitung, eds. H. Ueckert and D. Rhenius, pp. 153-161. Bern: Huber.

(6) Klix, F. 1971. Information and Verhalten. Stuttgart: Huber.

(7) Kluwe, R. 1979. Wissen und Denken. Stuttgart: Kohlhammer.

(8) Krause, B., and Krause, W. 1980. Human problem solving. In Psychological Research. Humboldt-University Berlin 1960-1980, eds. F. Klix and B. Krause, pp. 108-132. Berlin: VEB Deutscher Verlag der Wissenschaften.

(9) Lüer, G. 1973. Gesetzmäßige Denkabläufe beim Problemlösen. Weinheim: Beltz.

(10) Neisser, K. 1976. Cognition and Reality. San Francisco: Freeman.

(11) Newell, A., and Simon, H.A. 1972. Human Problem Solving. Englewood Cliffs: Prentice Hall.

(12) Nisbett, R.E., and DeCamp Wilson, T. 1977. Telling more than we can know: Verbal reports on mental processes. Psychol. Rev. 84: 231-259.

(13) Norman, D.A. 1980. Twelve issues for cognitive science. Cog. Sci. 4: 1-32.

(14) Putz-Osterloh, W. 1981. Problemlöseprozesse und Intelligenztestleistung. Bern: Huber, in press.

(15) Putz-Osterloh, W., and Lüer, G. 1981. Über die Vorhersagbarkeit komplexer Problemlöseleistungen durch Ergebnisse in einem Intelligenztest. Z. exp. a. Psy. 28: 309-334.

(16) Selz, O. 1913. Über die Gesetze des geordneten Denkverlaufs. Stuttgart: Spemann.

(17) Simon, H.A. 1979. Information processing models of cognition. Ann. Rev. Psychol. 30: 363-396.

(18) Sternberg, R.J. 1977. Intelligence, Information Processing, and Analogical Reasoning: The Componential Analysis of Human Abilities. New York: Erlbaum.

Animal Mind - Human Mind, ed. D.R. Griffin, pp. 177-200.
Dahlem Konferenzen 1982. Berlin, Heidelberg, New York: Springer-Verlag.

Ascent of Apes

D. J. Gillan
University of Pennsylvania Primate Facility
Honeybrook, PA 19344, USA

Abstract. Is reasoning a uniquely human ability? This article reviews experiments on reasoning in chimpanzees. In the first series of experiments, a 16 year old female chimpanzee, Sarah, demonstrated the ability to reason analogically in a variety of analogy problems. She chose the correct stimulus B' to complete an analogy A is to A' same as B is to B' and chose the correct predicates to complete analogies A is to A' same as B is to B' and A is to A' different from B is to C. The second series of experiments investigated transitive inference. A 6 year old female chimpanzee, Sadie, was trained on adjacent pairs from a linear series, F r E, E r D, D r C, C r B, and B r A, where the relation r was "more food than." Sadie then was given the novel nonadjacent pairs, EB, EC, and DB; she chose E and D indicating the ability to make transitive inferences. The third series of experiments placed four chimpanzees in a reasoning task similar to a problem likely to confront chimpanzees in their natural environment. A subject had to choose between two limited food sources. Just prior to making the choice, the chimpanzee received negative information from which it could infer that food was no longer in one of the locations. The chimpanzees demonstrated their inferential abilities by consistently going to the other location.

INTRODUCTION

What a piece of work is a man, how noble in reason, how infinite in faculties; in form and moving how express and admirable; in action how like an angel; in apprehension how like a god: the beauty of the world, the paragon of animals.

Hamlet, II,ii

An important function of both the biological and psychological sciences is to help humans understand their relation to the world. Accordingly, one of the central questions that these sciences have investigated is: What makes humans unique? Psychologists' answers to this question center around cognitive abilities that might be uniquely human. From the time of Descartes, one frequent answer to the question of cognitive abilities that are uniquely human has focused on reasoning (for examples, (8,9,10,12,20)). It is of historical importance that the scientific study of animal learning began when Morgan (12) and Thorndike (20), separately, attempted to investigate anecdotes reported by Romanes (16) and others which suggested that nonhuman animals could reason. Both Morgan and Thorndike explained these previous anecdotal data by associative mechanisms that did not require granting inferential abilities to non-humans. In the theories of both of these influential psychologists, reasoning was the principal cognitive difference between humans and other animals.

Given both the scientific and popular interest in differences between humans and nonhumans, it is surprising that virtually no experiments have investigated reasoning in nonhuman animals in the past fifty years. The bulk of the data, fifty years ago and up to the present, supported the Morgan-Thorndike view that nonhuman animals could not reason. It is likely that the paucity of data on animal reasoning is largely because human reasoning primarily involves language for its expression. For example, the common syllogism, "Socrates was a man; all men are mortal; therefore, Socrates was mortal," would be difficult, or impossible, to express without language. Thus, a researcher interested in reasoning by nonlinguistic animals is faced with two basic methodological problems: How to present reasoning problems to nonhuman subjects, and how to have these subjects answer the problems. There are two approaches to these problems. First, the researcher could try to teach the animals human (or human-like) language (for an example of such a language training program, see (14)). An alternative approach would be to present the information from reasoning problems without using language and set up the problems so that the subjects could give nonverbal answers. Although some of the work discussed in this paper used a language trained chimpanzee, Sarah (14), my colleagues, David Premack and Guy Woodruff, and I primarily used the second approach.

Our studies investigated reasoning in chimpanzees. Chimpanzees seemed to be good candidates for experiments on reasoning because of their conceptual abilities (e.g., (14)) and their problem-solving skills (e.g., (11,15)). We chose three types of reasoning problems to give to chimpanzees: (a) analogical reasoning, that is, judging the equivalence between two relations or completion of one relation so that it is equivalent to another relation; (b) transitive inference, that is "A r B, B r C, therefore A r C," where r is some transitive relation, such as greater than; (c) "natural" reasoning problems in which chimpanzees had to infer that food was no longer in a location on the basis of partial information. These types of reasoning were chosen because of their importance in both psychologists' and logicians' discussions of reasoning (1,4, 17-19), because they represent the two major classes of reasoning, induction (analogical reasoning) and deduction (transitive inference), and because they involve problems modeled after those commonly given to humans (analogical reasoning and transitive inference), as well as problems that a chimpanzee might face in its natural environment (the natural reasoning problems).

ANALOGICAL REASONING

The experiments on analogical reasoning used Sarah, a 16 year old female chimpanzee (Pan troglodytes), who previously had served in a variety of experiments on language and other cognitive processes (14,15). Analogy problems took one of two forms: (a) A is to A' is the same as B is to either B' or C (forced-choice problems); Sarah's task was to complete the analogy by choosing B', (b) A is to A' is either the same as or different from B is to B' or B is to C, respectively (same-different problems); she had to complete the analogy by choosing the correct predicate, either her symbol for "same," if the problem had A:A' and B:B', or "different," if the problem had A:A' and B:C. A more detailed description of the procedures is given in Gillan, Premack, and Woodruff (7).

The initial experiments examined Sarah's analogical reasoning abilities with geometric figures. These analogies involved perceptual relations among stimuli, e.g., a change in size, in color, or in marking (dot vs no dot). For example, a forced-choice problem given to Sarah had a medium-sized, blue, marked sawtooth-shaped stimulus as A and a medium-sized, blue, unmarked sawtooth shape as A'. Thus, the A-A' relation was the removal of a dot. The B stimulus in the problem was a large, orange, marked, crescent shape. The alternatives that Sarah received along with the problem were a large, orange, unmarked crescent shape and a large, blue, marked crescent shape. Because the relation between B and the unmarked crescent shape was the same as the A-A' relation, the unmarked alternative was the correct one, B'. Note, however, that the incorrect alternative, C, the blue, marked, crescent, also was closely related to B, the relation being a change in color. In fact, both alternatives, B' and C, differed from B in only one dimension. Thus, Sarah's choice of B' on this problem probably was not due simply to physical similarity of B' and B since C was also quite similar to B.

To control for the possibility that Sarah's choice on forced-choice analogy trials was determined by physical similarity with A' or B, we constructed all trials in pairs. Both trials of each pair had the same A' and B and the same alternatives; the stimulus that differed between trials in a pair was A. Consequently, the A-A' relation also differed between trials in a pair. On one trial of a pair, one alternative was correct; on the other trial, the other was correct. The correct alternative was always a function of the A-A' relation. Continuing the above example, the second trial in the pair had a medium-sized, orange, unmarked sawtooth shape (A), and a medium-sized, blue, unmarked sawtooth shape (A'); thus, the A-A' relation was a change in color. The B stimulus in the problem was, as before, a large, orange, marked crescent shape and the alternatives were, as before, a large, orange, unmarked crescent shape and a large, blue, marked crescent. However, unlike the previous trial in this pair, the latter alternative was correct because it now had the same relation to B that A' had to A.

Sarah chose B' on 44 of 52 (85%) of the forced-choice analogy problems constructed as described above. Because every problem was unique, Sarah could not have learned the correct answers to specific problems through trial and error. In addition, she chose correctly on the first four problems, which suggests that she possessed analogical reasoning processes prior to the start of this experiment. However, regardless of whether she acquired these processes on the initial trials of this experiment, acquired them due to her previous training (which did not include any experience with analogy problems), or was born with the ability to reason analogically, the data clearly indicate that apprehension of relations is a central process in Sarah's solution of analogy problems: Her choice of an alternative shifted as the A-A' relation (and the A-B relation) shifted. Thus a relation between two stimuli determined Sarah's choice. The ability to make one relation equivalent to another relation is commonly accepted as requiring analogical reasoning (17). Therefore, Sarah's performance on forced-choice figural analogy problems is strong evidence for analogical reasoning by a chimpanzee.

An additional defining characteristic of analogical reasoning is that it involves a judgment of equivalence of two relations. Accordingly, we gave Sarah figural analogy problems with A:A' ? B:B' or A:A' ? B:C in the problem display and gave her her symbols for "same" and "different" as alternatives. She correctly assigned the predicate "same" to A:A' ? B:B' on 13 of 18 (72%) of those trials and the predicate "different" to A:A' ? B:C on 13 of 18 (72%) of those trials. As in the initial experiment, all 36 problems were unique, so Sarah could not have learned the correct response to any problem through trial and error. These results clearly indicate that Sarah's analogical abilities are not restricted to forced-choice problems.

Sarah's correct choice of B' on forced-choice problems and of the predicate on same-different problems with figural analogies is clear evidence of reasoning by a chimpanzee. However, the perceptual relations among stimuli in those experiments differ from much human analogical reasoning which involves conceptual relations. For example, a conceptual analogy, such as, baseball is to diamond as basketball is to court, requires retrieval from long-term memory of attributes associated with baseball and diamond to determine their relation and retrieval of attributes associated with basketball and court to determine their relation. Because Sarah has only a small lexicon of symbols, we could not present her with symbolic conceptual analogies. Rather, we gave her forced-choice analogy problems in which the stimuli were common household objects with which she was familiar and the relations between the objects were conceptual. Those relations were either functional, e.g., lock is to key, or spatial, e.g., flower is to stem.

An example conceptual analogy problem given to Sarah was as follows: Paper (A) is to scissors (A') same as apple (B) is to either knife or plate. On this problem, Sarah correctly chose the alternative that had the same relation to B that A' had to A, the knife. However, such a choice could be explained by noninferential accounts, such as, an associative explanation. For instance, it might be that Sarah chose the knife because it was more strongly associated with an apple than the plate was associated with an apple. To control for the possibility that Sarah's choice was determined solely by an association with B, we again constructed the problems in pairs. Each problem in a pair had the same B stimulus and the same alternatives. In one problem of a pair one alternative was correct, in the other problem, the other alternative was correct. The correct alternative was a function of the A-A' relation, not an association with B. The second problem of the above example problem pair was lid (A) is to jar (A') same as apple (B) is to either knife or plate. On this problem, Sarah correctly chose the plate, B'.

Overall, Sarah chose correctly on 15 of 18 (84%) of the conceptual analogy problems. As before, the problems were unique, so Sarah could not have learned the answer to the problem through trial and error. In fact, because problems were constructed in pairs in which the correct alternative for one problem was the incorrect alternative for the other, had Sarah learned to choose a particular alternative in the first problem of a pair, she would have always been incorrect on the second problem of the pair. Rather, to choose the correct alternative consistently, Sarah had to apprehend the A-A' relation. That is what the data indicate that she did: When the A-A' relation changed, her choice changed accordingly.

The data from these experiments demonstrate that Sarah can reason analogically, that is, can both make judgments about the equivalence of two relations and complete one relation so that it is equivalent to another relation. It is notable that Sarah expressed her reasoning abilities in the present circumstances. Why did she not make choices based on physical similarity or associations with individual stimuli? It is possible that given a different problem display, perhaps one with the stimuli in different physical relations to one another, she would have expressed cognitive abilities other than reasoning. This suggests that the way a problem is presented may influence whether an organism accesses its reasoning abilities. This point may be applicable to tests of human reasoning, a central part of intelligence testing. It may be that differences among groups in the population in the ability to solve "culture-free" reasoning problems similar to the figural analogies given Sarah, e.g., Raven's progressive matrices, reflect not a difference in reasoning ability but a difference in the cognitive processes accessed by a given problem display. That is, certain people when confronted with a particular problem display may attempt to solve the problem either noninferentially or with reasoning processes not relevant to the problem. However, if presented with the same problem in a different physical display, those subjects might reason correctly. In addition, it may be that experience is important in determining which problem display will access which cognitive processes. Thus arguments about genetically-based group differences in human intelligence that are based on data from "culture-free" tests ignore the possibility of cultural influences on the strategies elicited by the physical structure of the problem.

Sarah's performance on analogy problems supports the proposition that analogical reasoning is a basic type of reasoning in chimpanzees as well as humans. These experiments do not indicate how widespread this reasoning ability is among primates, however. For example, Sarah's special training may have influenced her reasoning abilities. However, analogical reasoning requires two processes: apprehension of relations between stimuli and the ability to make a same-different judgment. Any species that can both apprehend relations and make same-different judgments can almost certainly combine these processes to make same-different judgments about relations, that is, can reason analogically.

What function might analogical reasoning serve for a chimpanzee? Analogical reasoning has been proposed as an important mechanism for human problem solving (5); it is likely that analogical reasoning also plays a role in chimpanzee problem solving. For example, when a chimpanzee is faced with a novel problem and several possible solutions, the animal may search through its memory for a similar problem. Once a memorial representation of a similar problem and its solution were found, the chimpanzee might determine what the relation between that previous problem and solution was. The animal might then determine the relations between the present problem and the various possible solutions. When it found a solution that had the same relation to the present problem as the relation between the previously experienced problem and solution, it would choose that solution as the one to perform. This informal model of chimpanzee problem solving suggests that chimpanzees' solutions of novel problems are based on their knowledge of the solution of previously experienced, similar problems. Likewise, failure to solve a problem may be due, at least in part, to either a lack of experience with similar problems or an inability to access the memorial representation of the correct problem. It should be noted that not all problems could be solved through analogical reasoning: the initial problems solved by a young chimpanzee could not evoke a memory of other similar problems since the young animal would have had no prior problem solving experience. Those initial problems might be solved by trial and error and then serve as a knowledge base for subsequent analogical problem solving.

The above informal model of chimpanzee problem solving suggests that a crucial factor in this use of analogical reasoning is the ability of the chimpanzee to access the relevant previous problem. This might occur in either of two ways: First, it is possible that the chimpanzee searches in memory for a problem that shares certain features with the current problem. For example, if the chimpanzee were faced with a problem of reaching bananas overhead, it might search its memory for a representation of a previous problem in which there were out-of-reach bananas or for a problem in which there was a desired object overhead. The second, and more complex, possibility is that the chimpanzee determines the relation between the current problem and one possible solution, then searches in memory for a representation of a previous problem and solution with that relation. Upon finding such a relation, the animal would then choose that solution to the current problem.

TRANSITIVE INFERENCE

A series of stimuli can be a transitive series, a nontransitive series, or an intransitive series. In a transitive series, if A r B and B r C, then A r C where r is some relation, e.g., greater than. In nontransitive and intransitive series, A r B and B r C do not necessarily lead to A r C. Transitive inference can be assumed if a subject is given only information about adjacent pairs in the series, e.g., A r B and B r C, and the subject deduces the relation between nonadjacent stimuli, e.g., A r C. Transitive inference has been an important and controversial type of reasoning studied in human children and adults (1,2,13). Accordingly, we studied transitive inference in the chimpanzee to discover whether that primate species could also make transitive inferences.

The subject in the transitive inference experiments was Sadie, a 6 year old female chimpanzee, who had not been trained in a language system. Sadie's lack of language provided an advantage - we could determine the importance of language for transitive inferences (3). However, it also made it necessary for us to present the information about relations between stimuli in a nonsymbolic way. To accomplish this, we gave Sadie pairs of adjacent stimuli from a series of relative food values: Stimulus E had more food than D, D more than C, C more than B, and B more than A. In the subsequent test, Sadie had to choose between the nonadjacent stimuli, D and B.

One way to present the above stimulus series, E>D, D>C, C>B, and B>A, would be to have E contain 4 pieces of food and D contain 3, D would contain 3 while C contained 2, C would contain 2 while B contained 1, and B would contain 1 while A contained none. Although this procedure would nonverbally represent the relation "more food than," it would not be satisfactory because it would not control for simple, noninferential explanations if Sadie later chose D over B on nonadjacent test trials. For example, one might reasonably assume that the 3 pieces of food paired with D in such a procedure would more strongly reinforce the choice response than would the 1 piece paired with B. Consequently, choice of D on DB trials could be interpreted as being due to reinforcement, not reasoning. We solved this problem by setting up trials so that, on every trial, the stimulus with the greater amount had one piece of food, and the stimulus with the lesser amount had none. Thus, the trials given to Sadie were E+D-, D+C-, C+B-, and B+A- where + indicates a piece of food and - indicates none. With this procedure, the average amount of food contained in E was 1, in D, C, and B was .5 and in A was 0. Thus, D, C, and B could be placed in a series only on the basis of their relations to one another and to E and A. Consequently, D could be chosen over B on the basis of knowledge of the series order, E>D>C>B>A, but not on the basis of differential reinforcement. Nor could D be chosen over B due to the order of presentation of adjacent pairs. The adjacent pairs, ED, DC, CB, and BA, were presented in an irregular order, rather than in their sequential order, during both training and tests. The procedures are repeated in full detail in the research report of these experiments (6).

Following extensive training on the adjacent pairs from the series, Sadie consistently chose the container with food on each pair. Then, she was tested with the nonadjacent pairs, as well as with the 4 adjacent pairs. On the test trials with the nonadjacent pair, DB, both nonadjacent stimuli contained food on 6 trials and neither contained food on 6 trials. The stimuli on trials with adjacent pairs contained food and no food as they had during training. Sadie's choice on adjacent pairs in the test continued at a highly accurate level: She chose the container with food on 32 of 36 trials (89%). Furthermore, she chose D over B on all 12 nonadjacent test trials (100%). Sadie has not been trained on DB, yet she chose D on DB trials from the start of the test. In addition, she continued to choose D even though it contained food on only half the DB trials. Her performance on this novel, nonadjacent pair is consistent with the hypothesis that she inferred the order of the transitive series.

In a second experiment, Sadie provided further evidence that she could make transitive inferences. She received training on the adjacent pairs, F+E-, E+D-, C+B-, and B+A-, and was tested on the nonadjacent pairs EB and EC, in addition to the adjacent pairs. She chose correctly on 35 of 40 (88%) of the adjacent pairs in the test and chose E on 10 of 12 EC and 10 of 12 EB trials (both 83%). Sadie chose E on the initial trials of both of these nonadjacent pairs. Her immediate choice of E in these novel, nonadjacent pairs and of D in DB trials of the previous experiment suggest that she did not simply learn to choose D or E during the test. Rather, the data suggest that she inferred the series order, then used that order to choose D and E on nonadjacent trials.

Sadie's choice of D and E on novel nonadjacent pairs DB, EB, and EC, was consistent with the hypothesis that she had inferred the series order, F>E>D>C>B>A, on the basis of experience with adjacent pairs and used that series order to choose on nonadjacent pairs. A prediction from the hypothesis that she had inferred the series order is that manipulation of that order should affect her choice on nonadjacent trials. One salient feature of the series order given to Sadie was that it was a linear order, with a high and a low end point, F and A, respectively. To manipulate the series order, we gave Sadie experience with the same adjacent pairs as before, F+E-, E+D-, D+C-, C+B-, B+A-, but added the pair A+F-. The addition of A+F- joined the two end points of the series and made it nonlinear. That is, there was no longer a unidimensional scale of relative food values underlying the stimuli F to A. Therefore, if a linear order were necessary for choice of E and D on nonadjacent trials, then Sadie would not consistently choose those stimuli following training on the nonlinear order. It is noteworthy that the only stimuli directly manipulated in this change in series order were F and A, not the stimuli to be tested, E, D, C, and B.

After Sadie had received extensive training on the pairs, F+E-, E+D-, D+C-, C+B-, B+A-, and A+F-, she chose the stimulus with food in it consistently. During the following test with nonadjacent pairs, DB, EB, and EC and adjacent pairs, Sadie chose correctly on 90% of the adjacent pair trials, a performance comparable to that in the previous linear order experiments. In contrast to the between-experiment consistency on adjacent pairs,

Sadie chose E and D on only 18 of 36 (50%) of the EB, EC, and BD trials. It is interesting to note that in the initial experiment, Sadie chose D on all of the DB test trials and, in the second experiment, she chose E on 20 of 24 EB and EC trials. The conditions of reward and nonreward on DB, EB, and EC trials were the same in the present experiment as in these previous experiments. A hypothesis that Sadie chose D and E in those earlier experiments because she had learned to do so during the test does not appear to be able to account for her failure to choose D and E consistently in the present test. The only condition that changed between the previous experiments and the present one was the series order. In the first two experiments, the series order was linear, and Sadie chose D and E. In the third experiment, the series order was nonlinear, and she chose inconsistently. In addition, a subsequent experiment showed that after reinstituting the linear order by presenting F+A- trials, as well as F+E-, E+D-, D+C-, C+B-, and B+A- trials during a retraining phase, Sadie again chose D and E consistently, on 30 of 36 (83%) of nonadjacent trials. She chose D on the initial DB trial and E on the initial EB and EC trials in this test. Thus the choice on nonadjacent trials was a function of the linearity of the stimulus order as determined by the relation between F and A. As noted above, this is of special interest because the stimuli manipulated to modify the linearity of the series, F and A, were not the test stimuli on nonadjacent trials; those test stimuli were E, D, C, and B.

The finding that the relation between F and A had a substantial impact on performance on nonadjacent trials further indicates that Sadie had a mental representation of the series order. That representation is notable because Sadie never received training with the entire series at the same time. Rather, she had to infer the series and its order on the basis of experience with only the adjacent pairs of stimuli.

Not only did Sadie infer the series order, but she used it to make choices on novel nonadjacent trials, much as a human would use a series F>E>D>C>B>A. However, it is likely that Sadie's appreciation of the relations between stimuli in these experiments was not identical to an adult human's appreciation of the relation "greater than." For example, imagine what a human's response would be if given adjacent pairs from a nonlinear series such as those given Sadie, F>E, E>D, D>C, C>B, B>A, and A>F. A human given such information might tell the experimenter "that's impossible". Sadie could not express any incredulity she might have felt. In fact, after many trials she learned to respond appropriately on each individual pair in the nonlinear series. Sadie's ability to learn the nonlinear series suggests that she may have approached the relation between adjacent stimuli in these experiments not as a relation of "greater than," but as an unknown relation. That is to say, she made no initial assumption that it was a transitive or an intransitive relation. However, if she did first approach the relation as indeterminant, her choice of D and E on novel, nonadjacent test trials indicates that, by the time of the tests in the initial experiments, she assumed that the relation was transitive. It seems likely that the basis of her assumption about the transitive nature of the relation was the linearity of the series order in those experiments. In fact, in all of the experiments, it appears that the structure of the series determined Sadie's appreciation of the relations among individual stimuli: When the series was linear, the relation was transitive; when the series was nonlinear, the relation was not transitive.

Would the series structure also determine a human's assumptions about transitivity and intransitivity of a relation? In an investigation of this question, I gave ten people the following story: "While at the zoo, you go to the ape house. You watch the chimpanzees playing and fighting. In the fights that you watch, Bonzo defeats Fred, Fred defeats Elmer, Elmer defeats Bert, Bert defeats Roscoe, and Roscoe defeats Bonzo." Then I asked the people who would win fights between various chimpanzees, both those who had fought each other in the story and novel pairings of chimpanzees. This nonlinear series is comparable to that given to Sadie. Unlike the procedure given Sadie, however, I allowed the human subjects to answer "Don't know." Five subjects said that they did not know who would win between Bert and Fred; the Bert vs. Fred pairing was comparable to the DB pair in Sadie's nonlinear series. When interrogated further these five said that they did not know who would win because the series was nonlinear. The remaining five subjects said that Fred would defeat Bert. Four of these five subjects said they did so because the series was linear with the exception of the fight between Roscoe and Bonzo which they treated as a fluke. The choice of Fred over Bert by these subjects suggests that many humans may attempt to impose a linear structure even on a nonlinear series.

It is likely that one use of transitive inference by chimpanzees is in social situations similar to those described in the above story given to the human subjects. For example, assume that chimpanzee B had lost a fight to chimpanzee C, and C had lost to chimpanzee D. Chimpanzee B would not have to fight and presumably lose to D if B could infer the series order D r C r B and assumed that r was a transitive relation (see also the chapter by Kummer in this volume). It may be that one of the environmental pressures for the evolution of the component processes that make up transitive inference came from social interactions. The fewer fights engaged in by an individual, the more time and energy he could devote to reproduction. Because chimpanzees with inferential abilities would not have to fight many other chimpanzees to determine their rank in the social hierarchy, they would have a reproductive advantage over chimpanzees without inferential processes. In addition, if social interactions were crucial for the evolution of transitive inference, then nonsocial primates may not have evolved this ability.

NATURAL REASONING

In the experiments on analogical reasoning and transitive inference, chimpanzees were given reasoning problems modeled after those typically given to humans. In constrast, the present section describes a series of reasoning problems modeled after a type of problem that chimpanzees might face in their natural environment. In these experiments, a subject chimpanzee stood in a central location and observed the baiting of two hidden locations in a familiar, walled field. Next, the chimpanzee received negative information about one of the two food locations. The type of negative information varied from experiment to experiment. Then, the subject chose to go to one location or the other. A similar problem would exist for chimpanzees in the wild when they had knowledge of two limited food sources, but received negative information about one. For example, one type of negative information in such a situation would involve observing other chimpanzees walking away from one food source with armloads of food.

In the initial experiment, each of 4 subjects received 12 trials in which he or she observed a second chimpanzee walk to one of the two hidden locations, pick up the food, and walk with it back to the central location. The location to which the second chimpanzee went was on the left side of the field on 6 trials and on the right side on 6 trials. The removal of the food constituted the negative information in this experiment. Then the subject chose between two food sources. On the surface, it might appear that the subject has direct knowledge of the absence of food in one location and, as a consequence, would not have to infer its absence. However, keep in mind that the food locations were hidden from the subject's view. Thus, the subject never directly saw that food was present or absent in a particular location. Lacking such direct perceptual information, the subject would have to make inferences about the presence of the food. Admittedly, the negative information in this first experiment provided the subject with a strong reason to infer that the food was no longer in one of the locations. In this experiment, the 4 subjects went to the location to which the other chimpanzee had not gone on 44 of 47 (94%) of the trials. All 4 subjects went to the food-containing location on both the initial trial when that location was on the right and the initial trial when it was on the left of the field. The accurate performance on the initial trials indicates that the subject chimpanzees did not acquire their response during the course of the experiment. Overall, Sadie and Bert were correct on all 12 trials, and Luvie and Jessie were correct on 10 of 12 and 10 of 11 trials, respectively.

These data suggest that chimpanzees can infer that two initially equal, limited food sources have become unequal when they see another chimpanzee forage at one of the food sources. By making such an inference, a chimpanzee could choose the more fruitful food source, even in the absence of direct perceptual information about the amount of food in the foraged location.

In subsequent experiments, we varied the type of negative information about the food location. In the second experiment, as in the first, the subject saw food placed in two locations and, eventually, chose to go to either Location A or Location B. However, after seeing the baiting, the subject was removed from the field. While the subject was gone, the second chimpanzee was taken to one location. The second chimpanzee, unseen by the subject, removed the food from that location, but held it in his or her hand rather than eating it. When the second chimpanzee had returned approximately one-half of the way back to the central location from that food source, the subject reentered the field to observe the second animal's return. Thus, the subject could see that the other chimpanzee was returning with food from a particular direction. Then, the subject chose between Location A and Location B. Surprisingly, the four chimpanzees did not choose to go to Location B consistently when they saw the other animal return from Location A with food or vice versa. Rather, they went to that location on only 26 of 48 (54%) of the trials. Apparently, observing the second chimpanzee return from one location carrying food was not adequate negative information.

One possible explanation of the above result is that seeing the second chimpanzee remove food from a location is necessary for the subject chimpanzee to infer the absence of food from that location. If so, it would indicate that chimpanzees' inferential ability in this situation is severely limited. However, the next experiment showed that this was not the case. In that experiment, the subject saw both locations baited and eventually had to choose to go to one or the other.

However, after the baiting and before the choice, the subject observed the second chimpanzee go approximately one-half of the way up the path to one location. On half of the trials, the subject observed the second chimpanzee go toward the food location on the right side of the field and on the other trials, observed the second chimpanzee go toward the food location on the left side. When the second animal was half-way to the food source, the subject was removed from the field. Then, after the second chimpanzee had removed the food and returned to the central location, the subject was brought back to the field, shown that the second animal had food, and allowed to choose between the two locations. Thus, as in the previous experiment, the subject had information about the direction taken by the second chimpanzee and that that animal had food, but the subject did not observe the removal of the food. However, unlike the second experiment in this series, the subject chimpanzees went to the correct location consistently, on 21 of 24 (88%) of the trials. Sadie and Bert went to the correct location on all 6 trials that they received; Luvie and Jessie were correct on 5 of 6 and 4 of 6 trials, respectively. In addition, all four were correct on the first trial. The results of this third experiment indicate that observation of the removal of food is not necessary for chimpanzees to infer that the food is gone from a location. Because the chimpanzees in this experiment never saw the second chimpanzee at the food location, but only saw that animal head in the direction of that location, consistently correct performance in this experiment required subjects to make a substantial inference that the food was gone.

It is notable that the subjects performed better when the two parts of the negative information about food in a location -

the direction of the second chimpanzee and that animal's possession of food - were temporally separate than when those pieces of information were temporally continguous. One explanation of the present findings is that chimpanzees can extrapolate forward to infer that another chimpanzee heading in one direction will continue in that direction toward some goal object. In contrast, chimpanzees may not be able to extrapolate backward to infer that another chimpanzee heading in one direction with a goal object is returning from a specific location where that object was.

A final experiment demonstrated that one chimpanzee, Sadie, did not need any directional information in this reasoning task. In this experiment, the subject saw both locations baited - Location A with an apple and Location B with an orange. Then, the subject was removed from the field. The second chimpanzee, unseen by the subject, went to one location and returned to the central starting point with the piece of fruit from that location. The subject was brought back to the field and shown that the other chimpanzee had a certain type of fruit. Then, the subject chose to go to Location A or Location B. The performance of two of the subjects was interesting - one because it displayed intelligence, the other because it displayed the opposite. Sadie chose the location with food on 10 of 12 (83%) of the trials. She was correct on the first left-going and the first right-going trials. Thus when given only the location of two different foods and the information that another chimpanzee had one of the foods, Sadie could consistently choose the location that still contained food. In contrast, Luvie chose the location with food on only 3 of 12 (25%) of the trials. She went to the location without food on the first 6 trials.

The data from Sadie and Luvie suggest that both chimpanzees had mental representations of the field and of the different pieces of fruit in each location. However, the marked difference in their behavior indicates that their use of that mental representation was very different. For example, Luvie's consistently incorrect choice might be explained by assuming that when she saw the fruit in the second chimpanzee's hand, it activated her mental representation of the location associated with that fruit. She then went in the direction of that location. Thus, this associative approach to the problem would have led Luvie to the incorrect location, the one lacking food. In contrast, it appears that seeing the fruit in the second chimpanzee's hand did provide negative information to Sadie. For example, when Sadie saw that fruit in the central location, it may have activated a negation rule, such as, "apple in central location implies no apple in Location A." Because Sadie had not learned a language system, her rule must have represented this information nonlinguistically. By applying such a negation rule, she could then accurately solve the problem and go to the location with food.

HOW NOBLE IN REASON

The prologue to this paper was a quotation from Hamlet which suggests that humans are the paragon of animals, at least in part, because of our reasoning abilities. Although the research described in the text that followed the quotation does not require that we revise Shakespeare's poetry, it does indicate that the ideas behind the poetry - ideas that have influenced our conception of the relation of humans to other animals since before the time of Shakespeare - should be revised. Humans are not unique in their ability to reason. Chimpanzees also possess a variety of reasoning abilities, both inductive and deductive.

The experiments described in this paper address an important question of the post-Darwinian era: Are certain cognitive abilities phylogenetically discontinuous? The results described here argue against discontinuity in reasoning abilities with a breakpoint at homo sapiens. Although it may be that only humans and chimpanzees can reason, such a statement is not warranted by any existing data. The question of which species other than humans and chimpanzees possess the ability to reason remains largely uninvestigated.

There are multiple approaches for the investigation of reasoning in animals. Our primary approach was to give chimpanzees problems that generally are considered to require reasoning when they are given to humans (2,19). We adopted more stringent criteria than most experiments on human reasoning by controlling for possible noninferential mechanisms for problem solution, such as, associations. Although our approach allowed us to demonstrate that reasoning is not unique to humans, it did not provide a general definition of reasoning itself.

I have suggested elsewhere that a general definition of reasoning may not be profitable, given the apparently disparate cognitive mechanisms of different types of reasoning (6). For example, the principal component processes of analogical reasoning are apprehension of relations and application of relations, as in a same-different judgment (e.g., (7,19)). In contrast, the processes involved in transitive inference are integration of adjacent pairs of stimuli into a linear series and use of that series to judge the relation between nonadjacent stimuli (e.g., (6,21)). At this point in our understanding of reasoning, it would take Procrustean means to force these two types of reasoning to fit the same process-oriented definition. Perhaps a more profitable approach for comparative psychologists and cognitive ethologists than worrying about definitions would be to determine the mental computations that various species can perform. For example, the present work demonstrates the chimpanzee's ability to apprehend both perceptual and conceptual relations, use relations in same-different judgments, construct an integrated series order on the basis of only pairs from the series, and so on. By focusing our research effort on the computational processes of animals, we ensure that as the definition of reasoning changes or even if, at some future time, the category of reasoning is abandoned by cognitive scientists we will still have valuable information about the cognitive abilities and limitations of organisms. That information might be used to relate ecological niches to those cognitive abilities in order to develop a theory of the evolution of mental processes.

Our knowledge of the cognitive abilities of apes, especially chimpanzees, has increased tremendously in the past fifteen years. With the increased knowledge has come an increased appreciation of the intellect of these primates. A variety of experiments on language, concept learning, classification, as well as those on reasoning reported here show us that the intellectual level of these animals is not as different from ours as was once believed. With each new experiment, the apes ascend in our estimation. However, these impressive results should not delude us into thinking that there are no cognitive differences between chimpanzees and humans. Several of the results from our experiments suggest differences between human and chimpanzee cognition. For example, in an analogy experiment, Sarah was given A:A' same as ? in the problem display and had both to choose the correct alternatives, B and B', from a set of six alternatives and to put B and B' in the correct order. Sarah correctly completed only one of 18 such problems. Humans undoubtedly would have much greater success with such problems. A second example of a likely difference between human and chimpanzee reasoning was evident in the second natural reasoning experiment. In that experiment, chimpanzees were shown another chimpanzee returning with the food from one of two known food locations, then chose between the two food locations. It is doubtful that humans would perform at chance in such a situation, as the chimpanzees did. Data such as these should make us realize that chimpanzees have evolved to meet the intellectual demands of an environment somewhat different from that of humans. Consequently, these apes may be noble in reason, but in apprehension are not identical to a human, much less like a god.

Acknowledgement. This research was supported by Grant BNS 77-16853 from the National Science Foundation and Grant 1-P01-HD-10965 from the National Institutes of Health to David Premack and a Cognitive Science Fellowship from the Sloan Foundation to the author. The author is now at General Foods Technical Center, Tarrytown, NY 10591, USA.

REFERENCES

(1) Bryant, P.B. 1974. Perception and Understanding in Young Children. London: Methuen.

(2) Bryant, P.B., and Trabasso, T. 1971. Transitive inferences and memory in young children. Nature 232: 456-458.

(3) Clark, H.H. 1969. Linguistic processes in deductive reasoning. Psychol. Rev. 76: 387-404.

(4) Copi, I.M. 1972. Introduction to Logic. New York: Macmillan.

(5) Gick, M.L., and Holyoak, K.J. 1980. Analogical problem solving. Cog. Psychol. 12: 306-355.

(6) Gillan, D.J. 1981. Reasoning in the chimpanzee: II. Transitive inference. J. Exp. Psychol. Anim. Behav. Proc. 7: 150-164.

(7) Gillan, D.J.; Premack, D.; and Woodruff, G. 1981. Reasoning in the chimpanzee: I. Analogical reasoning. J. Exp. Psychol. Anim. Behav. Proc. 7: 1-17.

(8) Hobhouse, L.T. 1901. Mind in Evolution. New York: Macmillan

(9) Huxley, T.H. 1897. Methods and Results: Essays. New York: Appleton.

(10) James, W. 1890. The Principles of Psychology, vol. 2. New York: Holt.

(11) Köhler, W. 1925. The Mentality of Apes. New York: Harcourt Brace.

(12) Morgan, C.L. 1894. Introduction to Comparative Psychology. London: Walter Scott.

(13) Piaget, J.; Inhelder, B.; and Szeminska, A. 1960. The Child's Conception of Geometry. London: Routlege and Kegan Paul.

(14) Premack, D. 1976. Intelligence in Ape and Man. Hillsdale, NJ: Erlbaum.

(15) Premack, D., and Woodruff, G. 1978. Problem-solving in the chimpanzee: Test for comprehension. Science 202: 532-535.

(16) Romanes, G.S. 1883. Mental Evolution in Animals. London: Kegan, Paul, Trench.

(17) Rumelhart, D.E., and Abrahamson, A.A. 1973. A model for analogical reasoning. Cog. Psychol. 5: 1-28.

(18) Spearman, C. 1923. The Nature of "Intelligence" and the Principles of Cognition. New York: Wiley.

(19) Sternberg, R.J. 1977. Intelligence, Information Processing, and Analogical Reasoning. Hillsdale, NJ: Erlbaum.

(20) Thorndike, E.L. 1898. Animal intelligence: An experimental study of the associative processes in animals. Psychol. Monog. 2(4, Whole No. 8).

(21) Trabasso, T.; Riley, C.A.; and Wilson, E.G. The representation of linear order and spatial strategies in reasoning: A developmental study. In Reasoning: Representation and Process, ed. R.J. Falmagne, pp. 201-229. Hillsdale, NJ: Erlbaum.

Animal Mind - Human Mind, ed. D.R. Griffin, pp. 201-224.
Dahlem Konferenzen 1982. Berlin, Heidelberg, New York: Springer-Verlag.

Cognitive Knowledge and Executive Control: Metacognition

R. H. Kluwe
Hochschule der Bundeswehr, Fachbereich Pädagogik
2000 Hamburg 70, F. R. Germany

Abstract. The domain of psychological research with regard to metacognition is discussed on the basis of a distinction between declarative and procedural knowledge in information processing systems. According to these types of knowledge, one can distinguish between a person's cognitive knowledge and executive processes. Cognitive knowledge refers to a person's stored information about human thinking, especially about the features of his own thinking. Executive processes refer to cognitive activity directed at the monitoring of the application and the effects of solution strategies and at the regulation of the course of one's own thinking. The subject of psychological research of metacognition is considered to be the control of activity in information processing systems.

INTRODUCTION

Metacognition refers to aspects of human thinking that have been emphasized as one field of psychological research essentially by John Flavell. He may be considered to be the pioneer in this field. According to his point of view, metacognition can be understood as cognition about cognition.

Since 1971 when Flavell raised the question: "What is memory development the development of?", an immense body of research has been published with respect to the development of "metamemory." The question has been answered by Flavell himself in a nearly programmatic manner: the subject of psychological research directed at memory development is not only the growing repertoire of memory strategies, but also the developing ability to monitor the application and the effects of memorization strategies as well as the knowledge about such strategies (7). Flavell's concept "metamemory" refers to this ability.

There is a tendency in the present psychological literature to create more "metas" like "meta-listening," "meta-persuasion," or "meta-communication." The usefulness of such inventions might be questionable. However, there are general attributes which are common to these activities referred to as "metacognitive": (a) the thinking subject has some knowledge about his own thinking and that of other persons; (b) the thinking subject may monitor and regulate the course of his own thinking, i.e., may act as the causal agent of his own thinking.

An example for the first category, usually referred to as metacognitive knowledge, is an individual's knowledge about his shortcomings in memorization or the strengths with regard to solving dynamic and complex problems.

The second category refers to an individual's cognitive activity having as its object his own cognitive enterprise, aiming at efficient and appropriate thinking. An example is changing the speed of information processing under time pressure, or the allocation of one's processing resources in order to focus on the relevant features of a situation. This second aspect of metacognition has not always been distinguished from individual knowledge about cognition. Flavell (8) refers to this aspect as "metacognitive strategies," Brown (2) uses the concept "metacognitive skills." I will discuss these two aspects of metacognition in detail later.

I found it useful here to look at psychological research in the field of metacognition, primarily developmental research in terms of a distinction made earlier by Ryle (28). According to his theoretical analysis, one may distinguish between declarative and procedural knowledge in information processing systems. Metacognitive knowledge as studied in developmental psychology is related to declarative knowledge stored in a system, i.e., stored data in long-term memory. Metacognitive strategies or skills, however, have to do with procedural knowledge, i.e., the stored processes of a system.

I will organize my discussion of psychological research on metacognition according to these two categories. Figure 1 illustrates my general understanding of the problem. My goal is to provide the reader with a broader description of "metacognition" as it is available from developmental research on "metamemory."

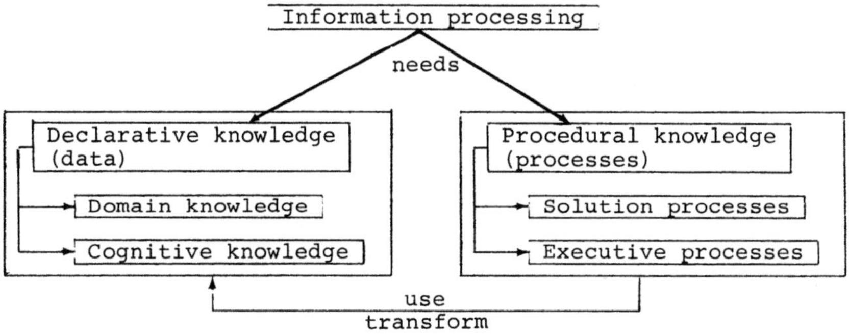

FIG. 1 - Types of knowledge necessary for problem solving.

Information processing requires data and processes. With respect to data I distinguish between domain knowledge and cognitive knowledge. By domain knowledge I mean an individual's stored information about the domains of reality, e.g., physics, politics, or sport. There is, for example, educational knowledge or physical knowledge. Cognitive knowledge refers to an individual's stored assumptions, hypotheses, and beliefs about thinking.

One can distinguish different categories of cognitive knowledge which will be described in more detail in the next part of this paper. I assume that there is cognitive knowledge invariant over domains, for example, the conviction that all difficult problems need intense concentration and the activation of all resources in order to come up with a solution. In addition there is domain-specific cognitive knowledge. An individual might, for example, be convinced that he/she is a poor problem solver with respect to physics problems, not, however, with respect to social problems. My assumption is that in solving a problem both types of data are necessary, i.e., data about the domain at hand and data about cognitive enterprises in general, as well as concerning this domain, are processed. What is called "metacognitive" knowledge in developmental psychological research corresponds to what I call cognitive knowledge here.

With respect to stored procedural knowledge, I assume that there are solution processes and executive processes. Solution processes are directed at the solution of a problem; executive processes monitor the selection and application as well as the effects of solution processes and regulate the stream of solution activity. The same is true for processes like communication. General solution processes have been described by Dörner (4), Lüer (19), and Newell and Simon (22). To the best of my knowledge the role of executive processes, however, has not attracted too much attention and has been limited in information processing models. I will discuss different types of executive decisions later.

A CLASSIFICATION OF COGNITIVE KNOWLEDGE

Flavell and Wellman (12) were the first to classify an individual's possible knowledge about memory. They distinguish between an individual's knowledge about subject, task, and strategy variables influencing memory processes and performance.

I proposed a classification of cognitive knowledge which is somewhat different (16). The starting point for this taxonomy is the assumption that stored knowledge about cognition does not differ from other stored information about the domains of reality. Cognitive knowledge is stored, organized, and processed in the same way as other knowledge about domains of reality. As a consequence, for example, the notation of semantic network models might be used to describe cognitive knowledge. However, developmental psychological research does not use this possibility as a framework for studying the development of cognitive knowledge. According to my viewpoint, there is no reason to use the concept "metacognitive" here to talk about this knowledge.

With respect to knowledge about domains of reality one can distinguish between (a) knowledge about subjects, objects, situations, or events in a domain, and (b) knowledge about means to transform such subjects, objects, or events. The same is true for the domain of cognition: An individual has some knowledge (a) about cognitive states and activities, cognitive processes, tasks, and performance, and (b) about ways to transform and change cognitive states, activities, etc.

With respect to these two basic categories one can again distinguish between (c) general and (d,e) diagnostic cognitive knowledge. General cognitive knowledge refers to a person's beliefs about the organization of the human information processing system (For an example see the assumptions of an 11-year old girl about the organization of the human memory in (14)). Diagnostic cognitive knowledge comprehends a person's knowledge, beliefs, assumptions, or hypotheses about features and traits of (d) his own and (e) other persons' thinking. Diagnostic cognitive knowledge about features of the individual's own cognitive states and activities refers to intra-individual differences. Diagnostic cognitive knowledge about the features of other persons' thinking refers to interindividual differences in cognition. An example is the conviction that individual A is a better problem solver than B.

As mentioned earlier these two categories of cognitive knowledge, i.e., general and diagnostic cognitive knowledge, presumably include knowledge which is invariant over domains and knowledge which is domain-specific. By combination of categories a) and b) with categories c) - e), six cells result (Table 1). Each cell corresponds to a certain type of cognitive knowledge.

A recent analysis of research in developmental psychology dealing with cognitive knowledge (16) shows that it is possible to classify most of the studies according to these six categories of cognitive knowledge. Most studies focus on developmental problems and are concerned with general cognitive knowledge - here again with respect to memory tasks. Only a few studies, however, deal with the knowledge as described in rows 2 and 3 of Table 1, indicating diagnostic cognitive knowledge. This is surprising since knowledge about one's own cognition might be considered to be at the core of cognitive knowledge and to be essential for coping with cognitive requirements. Bandura's new concept of self-efficacy (1), explanations of individual differences in cognitive performance, refers to a large extent to diagnostic cognitive knowledge. I will not repeat here the results of the numerous developmental studies in this field of research. Instead I will raise some issues that have to be taken into account in order to make corresponding psychological research more promising:

(a) Domain-specific cognitive knowledge. An individual's cognitive knowledge, especially about the features of one's own thinking, varies with respect to accuracy and correctness depending on one's knowledge and experience in a domain of reality. A physicist, for example, may hold quite accurate convictions about his own cognitive strengths and weaknesses in physics (about what he knows or does not know, about what he can solve with effort or what he can solve easily). This may also be true for his judgments about the cognitive traits of other persons working in this field. The same person's

TABLE 1 - A classification of cognitive knowledge

	KNOWLEDGE ABOUT COGNITIVE SUBJECTS AND AFFAIRS: COMPONENTS, STATES, AND ACTIVITIES OF THE COGNITIVE SYSTEM (a)	KNOWLEDGE ABOUT MEANS TO TRANSFORM COGNITIVE SUBJECTS AND AFFAIRS: TRANSFORMATION OF COGNITIVE STATES AND OF COGNITIVE ACTIVITIES (b)
GENERAL COGNITIVE KNOWLEDGE (c)	Knowledge about the general structure, organization, function of information processing; naive theories about memory and thinking; knowledge about states, e.g., to have a problem; knowledge about processes, e.g., learning, retrieval.	Knowledge about means, ways to transform cognitive states and activities; knowledge, e.g., that frequent rehearsal is a good memorization strategy, that repeated problem solving improves thinking, etc.
DIAGNOSTIC COGNITIVE KNOWLEDGE ABOUT ONESELF (d)	Knowledge about strengths, weaknesses, preferences referring to one's own cognitive functioning; knowledge about intraindividual variation of one's own thinking.	Knowledge about means, ways to transform one's own cognitive states and cognitive activities; knowledge about appropriate ways to meet cognitive requirements.
DIAGNOSTIC COGNITIVE KNOWLEDGE ABOUT OTHER INDIVIDUALS (e)	Knowledge about features of the cognitive states and activities of other individuals, about interindividual differences in cognition, as well as about intraindividual differences of other individuals in cognition.	Knowledge about the means, ways to transform the cognitive states and cognitive activities of other individuals.

cognitive knowledge and judgments will be less reliable and less accurate with respect to another domain, say politics. What I assume here is that an individual's cognitive knowledge, i.e., amount, accuracy, and correctness is probably dependent on one's experience with problem situations in a given domain of reality.

(b) Acquisition of cognitive knowledge. As mentioned above, I assume that there is no difference between storage, organization, and retrieval of cognitive knowledge compared to physical or economical knowledge. However, there remains one important difference: How do we acquire cognitive knowledge? Such knowledge is rarely published and is not taught explicitly. An individual is seldom provided with data about his own intellectual functioning. Of course, influences (e.g., parental teaching) can be considered to be sources of cognitive knowledge. However, this is not sufficient to explain how an individual develops a cognitive self-schema and a belief-system with regard to human information processing. We do not know, for example, what information of a problem solving process is stored as cognitive knowledge for possible future cognitive enterprises. We do not know how information is extracted from such a volatile process as one's own thinking, or how this information is integrated in a cognitive self-schema.

(c) Cognitive self-schemata. I propose to describe diagnostic cognitive knowledge in terms of schemata. This concept is familiar from semantic network models and would be useful as a framework within which to study diagnostic cognitive knowledge, e.g., in terms of cognitive self-schemata. Psychological research on metacognition would then become more than the mere collection of intuitions and assumptions about one's own and others' cognitive states and activities. The advantage of the schema-theoretical approach would be the emphasis on the functions of subjective cognitive knowledge when an individual has to meet cognitive requirements. In semantic network models of memory, schemata are related with certain assumptions about the processing, i.e., storage, organization, and retrieval of information.

(d) The functions of cognitive knowledge. What are the functions of cognitive knowledge? Does a problem solver need cognitive knowledge? Do developmental and interindividual differences with respect to amount, accuracy, correctness, and type of cognitive knowledge go along with differences in information processing and performance? It is still unknown whether, e.g., the knowledge of good problem solvers about cognition differs from that of poor problem solvers. As a consequence we do not know if it is useful and necessary to provide subjects with cognitive knowledge and to provide a person with data about his own intellectual functioning, its weaknesses and strengths. The hypothesis that there is a causal connection between cognitive knowledge and cognitive performance sounds plausible; but it cannot yet be supported by adequate evidence.

EXECUTIVE PROCESSES

Metacognition is a process; it is cognitive activity directed at the individual's own cognitive enterprise. I emphasize this procedural aspect which has been widely neglected thus far. Flavell (8) uses the concept of "metacognitive strategies" in order to describe what I believe is one function of executive processes: the monitoring of one's own thinking. A good example is the student who successively checks his cognitive state in order to keep track of his learning progress. Brown (2) speaks of "metacognitive skills" which are assumed to constitute the monitoring and the regulatory activity. She gives a more detailed description of such skills as checking, planning, or monitoring one's own thinking. Recently Sternberg (31) extended his theory of intellectual components by introducing so-called "metacomponents." Sternberg refers at this point explicitly to the work of Brown. The reason for this extension is individual differences in information processing which have to be explained. There are, for example, differences with regard to the duration of the application of cognitive operations in analogy tasks. Psychological studies dealing with such cognitive activities may be categorized in three groups:

(a) Instruction to monitor one's own cognitive states. Flavell, Friedrichs, and Hoyt (11) elaborated an experimental paradigm for the analysis of the developmental course of checking test readiness. The child has to memorize a set of items; the instruction requires the child to inform the experimenter when he/she believes that all items are well stored, and recall will be perfect. The dependent variable is the accuracy of the child's monitoring, i.e., checking and prediction of his own memory state.

(b) Monitoring <u>and</u> regulation. Studies of this type focus on the available information about the course of one's own thinking and its subsequent application to its organization. Masur, McIntyre, and Flavell (20) initiated such experiments, again using recall tasks. The authors report that even younger children acquire some information about the state of their own memories. They are able to identify those items which are well stored and those items which are not yet stored in memory for recall. However, unlike older children, they do not use this information that indicates monitoring activity for the organization of the subsequent memorization process. Older children regulate their own information processing according to this acquired information about the state of their own memories. They focus on those items that are not yet well stored. The relation between acquired information about one's own cognitive states and the rules for its subsequent use in the regulation of thinking have yet to be studied.

(c) Cognitive flexibility. Studies of this type require the regulation of one's own thinking processes in order to cope with changing situational demands. An early example is a study of executive functions published by Butterfield and Belmont (3). They examined the regulation of memorization processes by children and adults under changing conditions. The results of my own studies show that even 4-year old children are able to regulate their own problem solving processes in order to meet changing demands. The problem solving conditions in this experiment have been changed from reversibility to irreversibility of solution tactics (17).

The studies in groups (b) and (c) show what I believe is at the core of metacognition: cognitive activity directed at the acquisition of information about one's own cognitive enterprise (an expression which I borrowed from John Flavell), and the efficient regulation of one's own information processing according to the perceived demands. The consequence of this developing ability may be illustrated with the following analogy: children sometimes use tools, e.g., a hammer, a saw, or scissors. In the beginning, the application of such tools is often inadequate and clumsy. With more experience the tools are used better and more skillfully, i.e., the application is more tuned to the requirements of the material and the goal. The same may be true, according to my understanding of cognitive development, for cognitive tasks: children use their cognitive "tools" more flexibly and more efficiently; the organization of the course of their thinking corresponds increasingly well to the perceived cognitive requirements of a task. The underlying developing ability is the monitoring and regulation of their own thinking.

Most of the developmental studies in this field do not use the theoretical framework provided by information processing models of problem solving and thinking. It is, however, possible to locate the activity, described as monitoring and regulation, in information processing models. The basic idea is given with a computer simulation model of information processing. According to this model one can distinguish a main program (or executive program) and subroutines (21). This is in accord with my distinction made in Fig. 1 between executive processes and solution processes.

I assume that processes of monitoring and regulation of the course of one's thinking are a function of the executive component in information processing systems (see also (2,28)). The important difference between solution processes and executive processes is that the latter are not actually involved in solving the problem. They are necessary to monitor the effects of the solution activity and to regulate the organization and the course of the solution process. As Reitman once put it: the executive itself does not dirty its hands (27). Figure 2 shows a possible model of executive activity. According to this model, I distinguish between two types of procedural executive knowledge: (a) executive activity directed at the acquisition of information about the person's thinking processes, i.e., monitoring processes, and (b) executive activity directed at the regulation of the course of one's own thinking, i.e., regulation processes.

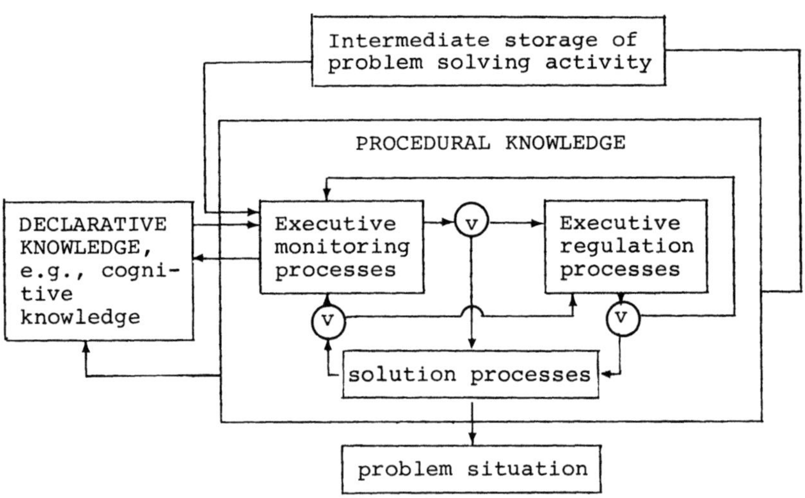

FIG. 2 - Hypothetical model of executive activity in problem solving.

This is somewhat different from usual conceptions concerning the executive component in information processing systems. As far as I know the executive program in computer simulation models is a more passive agent that depends on "bottom-up" information, i.e., results of the solution processes (subroutines) are given back to the executive component for further decision making. The executive program in the hypothetical model in Fig. 2 is assumed to be a more active agent or program that may acquire information about the states of the solution process actively and at any time, if necessary by monitoring processes. The acquired information is the basis for subsequent changes by regulation processes. This does not mean, however, that the monitoring processes accompanying the solution process are active with the same intensity throughout the whole problem solving procedure. There are probably certain phases with more automatic processing; however, we do not know the rules for the variation of monitoring activity in problem solving.

With respect to each of both categories of executive activity, it is possible to describe certain types of monitoring and regulatory decisions. Table 2 contains for the monitoring category four executive decisions which are assumed to provide the system with information about the present state of the cognitive enterprise. With respect to executive regulation there are four executive decisions aiming at the regulation of the speed, the subject, the intensity, and the resources of information processing. It is possible to group the developmental studies on the basis of this classification (17). I found, however, that it is impossible to get an overall picture of the developmental course of the ability to monitor and to regulate one's own thinking. The reason is that there are numerous studies dealing with different executive decisions with regard to different cognitive tasks.

There is some evidence for this classification from thinking aloud protocols (16). I analyzed such protocols by asking subjects to mark those statements that refer to their own solution approaches and to their own cognitive states. In one case I used a protocol which was available from Newell and Simon (22). It was surprising that approximately 60% of these statements, judged as "meta-statements," were not coded by Newell and Simon.

TABLE 2 - Types of executive decisions

EXECUTIVE MONITORING	EXECUTIVE REGULATION
1. Identification (What am I doing?),	1. Regulation of resources (Allocation of resources),
2. Checking (Did I succeed? Do I make progress?),	2. Regulation of the subject (What should I work on?),
3. Evaluation (Are there better alternatives? Is my plan good?),	3. Regulation of the intensity (amount of information that is processed; duration and persistence of processing),
4. Prediction (What could I do? What will result?)	4. Regulation of the speed (skip or add steps in problem solving), of information processing

In recent studies on problem solving, "meta-statements" from thinking aloud protocols have acquired more attention. Simon and Simon (30), for example, refer to such statements when analyzing the differences between experts and novices in problem solving.

It was of interest to me that Selz (29) in his early work on problem solving described processes that are quite similar to my distinctions with regard to executive activity. Selz discusses processes that initiate a solution process and that accompany this process. Among others he talks about monitoring processes ("Kontrollprozesse"), e.g., evaluation of strategies and pondering alternatives.

From studies with adults there is some evidence for interindividual differences in cognitive performance covarying with differences in executive activities as indicated by corresponding statements in thinking aloud protocols. Thorndyke and Stasz (32), for example, report such differences with regard to poor and successful learning processes. Obviously this is also true for problem solving. Dörner, Kreuzig, and Stäudel (6) found that good problem solvers reflect and monitor their own problem solving process more often than poor problem solvers. In both studies the authors rely on "metastatements" from thinking aloud protocols.

SOME EVIDENCE FOR COGNITIVE KNOWLEDGE AND EXECUTIVE CONTROL FROM ANIMAL PSYCHOLOGY

According to my understanding of metacognition as I have outlined here, we have to ask whether any animals show behavior that allows the inference that they have cognitive knowledge available, and that they are able to control and regulate such processes as their own problem solving or communication. Obviously there is no simple "yes" or "no" answer. Unfortunately, psychological research dealing with the development of metacognition does not give many hints for corresponding research with animals.

However, there are experiments with animals that show impressive results indicating some evidence for the availability of cognitive knowledge as well as of executive control. I will first discuss some of these results, and then add some remarks referring to the empirical research strategy.

Premack and Woodruff (23,24) published the results of careful experiments with chimpanzees that show (a) that chimpanzees are able to identify an event as a problem, and (b) that chimpanzees infer an intention or purpose of an agent. The basic procedure in these experiments has been to watch videotapes showing a human agent struggling with a problem. The chimpanzee then had to choose among alternatives, one of them showing the right solution step for the problem at hand. By varying the context of the problem-situations the authors could clearly show that the chimpanzee's concept of a problem is rather broad than narrow.

Premack and Woodruff infer from their observations that the chimpanzees impute mental states to others, i.e., that the animals have some theory of mind.

The importance of these results is that such a system of inferences about mental states of oneself and others allows the derivation of predictions. In research on the developmental psychology of metacognition this would have to be considered as "metacognitive knowledge" of an individual.

The experimental situation used by Premack and Woodruff furthermore requires a great deal of what Flavell in his earlier approaches to metacognition called "sensitivity" (12). It is the knowledge that certain situations do call for intentional problem-related behavior, that situations like those videotaped scenes require doing something special and require certain effort.

According to my taxonomy of cognitive knowledge, I would tend to consider the identification of a situation as a problem to be a basic element of an individual's cognitive knowledge. The most striking phenomenon of this experiment is that the chimpanzees obviously make inferences about intentions or purposes of the agent. Premack and Woodruff call this motivational inference, i.e., inference about motivational components of the situation. Inferences, however, about the knowledge of an agent, for example, about a child's knowledge in a problem-situation compared to an adult's knowledge, seem at least to be difficult for chimpanzees. The authors consider this an open question. Knowledge about another individual's knowledge would indeed be an impressive cognitive accomplishment. The distinction between inferences about the motivation compared to inferences about the knowledge of another individual as well as the assumptions about the difficulty of such inferences are highly interesting. As far as I know this has not yet been studied in developmental psychology.

However, the fact of inferences about motivational states alone is intriguing and corresponds to what Flavell (9) calls "psychological knowledge." Flavell recently assumed that individuals acquire such knowledge, that is, knowledge about psychological facts and states. Cognitive knowledge as described in Table 1 is only a small part of this domain of knowledge.

It would be useful to have more experiments of this type with very young children. Furthermore, one could study the chimpanzee's behavior under different conditions. For example, one could provide the animal with alternatives that do not include any reasonable solution for the problem. Does the chimpanzee show any response indicating puzzlement? Flavell uses the concept of "metacognitive experiences" in order to describe such states of an individual (10). Furthermore, one could provide the individual with different possible solutions that differ with respect to efficiency. We do not know if the chimpanzees would make such fine distinctions, e.g., would show some knowledge about the appropriateness of different solution approaches.

In a second extensive experiment Woodruff and Premack (34) examine the chimpanzee's ability in intentional communication. Again this is an extremely careful experiment. The procedure requires the animal to convey intentionally appropriate cues leading to a human subject's search for food. The experimental situation requires inferences about the intentions of the recipient of the information conveyed. A cooperative recipient is provided with correct information about the hidden food; the competing recipient is provided with misleading cues. In both situations the chimpanzees performance is appropriate. This requires abilities that can be considered as "metacognitive": (a) the chimpanzees infer purposes or motives from the recipient's behavior, as shown by the first experiment where a system of such inferences is considered to constitute a theory of mind. (b) The chimpanzees deliberately use means to convey either accurate or misleading information in order to support or prevent successful search for food. (c) The chimpanzees have some knowledge about the recipient's perception of their own behavior, i.e., about the recipient's inferences about the information they want to convey. (d) The individuals show sensitivity for the different communication situations, that is, the different recipients. They adjust their communication by changing the content of the conveyed information leading the recipient's search. They are able to control deliberately and regulate the information conveyed during communication.

Chimpanzees communicate incidentally a great deal. However, in terms of their intentional communication there is little research. Here we have carefully obtained results that lead us to believe that chimpanzees learn to be attuned to and responsive to those occasions when it is adaptive to try to communicate something immediately, e.g., they deliberately try to convey specific information.

In a recent chapter on monitoring social-cognitive enterprises, Flavell (10) proposed a broad description of this domain: "...what is thought about during a social-cognitive enterprise could be a perception, feeling, motive, ability, intention, purpose, interest, attitude, thought, belief, personality structure, or any other such process or property of self or other(s)" (p. 1). Premack and Woodruff provide us with both experiments with some evidence for such abilities in chimpanzees. With respect to intentional communication, however, we need more information about their ability to monitor the process of communication. "Monitoring a social-cognitive enterprise roughly means keeping track of how it is going and taking appropriate measures whenever it needs to go differently" ((10), p. 21). The experiment of Woodruff and Premack (34) has not been designed to study this question explicitly. It would require changing demands during communication with a recipient, for example, a competing recipient changing his mind and acting cooperatively.

A few remarks referring to research in this domain should be added. First, before designing comparative experiments, research in the sense of animal psychology would seem appropriate (see for this distinction (14)). It prevents wrong comparisons, clarifies the subject of comparison, and most important, permits the study of anecdotes, taking into account the diversity of animals.

Second, careful experiments like those done by Premack and Woodruff are necessary to avoid intriguing but misleading inferences from field observations. The hunting behavior of wolves might easily be called planful and deliberately strategic. However, we have no knowledge about the generality of such behavior; we do not know if wolves would react to varying behavior of the hunted animal in a way that executive control over their own strategic efforts could be inferred.

Third, further experiments dealing with metacognitive aspects of animals' problem solving or communication should take into account what I call the domain-specifity of cognitive knowledge. For example, animals living in social groups, e.g., chimpanzees, may develop some knowledge and executive control about communication. This is less likely the case with animals living in isolation. In a similar way, animals used to hunting, e.g., wolves, probably develop some specific knowledge and executive control about the organization and success of hunting compared to animals that need not hunt for food.

Fourth, before accepting such "ambitious" explanations as the availability of cognitive knowledge or executive control, one should look for simpler explanations. Premack and Woodruff in their experiments demonstrate this method in a very clear way. For example, in the case of intentional communication they ponder traditional learning principles as possible interpretations of their results (34). Contrary to the assumed control of the chimpanzee over its communication, the authors evaluate the possibility of learned stimulus-response associations. This procedure is also applied in the study on problem-comprehension (23,24). It seems to me that developmental psychological research on metacognition sometimes fails to exercise this reasonable caution.

CONCLUSION

According to my viewpoint the subject of research on metacognition is the control of activity in information processing systems. Pylyshyn (25) discusses this problem with respect to computer simulation models of information processing. This problem has already been well-known for years. Pylyshyn uses an analogy to illustrate the problem at hand: in human organizations the leadership and control is never given over to a "local expert" completely. The "superior authority" (the executive program) selects experts for certain tasks according to their qualifications (activation of subroutines, evaluation of strategies). Furthermore, the work of the members is monitored by some device, e.g., progress reports that have to be submitted. Pylyshyn emphasizes that the modelling of such monitoring and regulatory activity is a difficult problem. The main question here is not that of parallel processing, but that of the information which is exchanged between executive program and routines.

A special case is given when, in terms of Pylyshyn's analogy, the "superior authority" of an organization not only monitors and regulates the activity of the experts of the organization, but also makes its own work subject to monitoring and regulation. In Fig. 2 this possibility is taken into account by the intermediate storage of problem solving activity. That is, this process requires a type protocol of the traces of one's own cognitive activity (see also (5)). This protocol, which may vary with regard to its accuracy and completeness, can be the subject of the system's information processing. In the German problem solving literature this process is referred to as "self-reflection." To date I have not been aware of a solution to the problem of monitoring and regulation of one's own thinking in cognitive psychology (see (25)). However, I believe that it is necessary to extend existing models and to take into account the self-regulatory potential of problem solvers. There are several reasons why the information processing system needs such a self-regulatory mechanism:

(a) Human information processing has to meet varying cognitive demands. It is not always necessary to think hard, it is sometimes necessary to process information rapidly, etc. As a consequence the efficiency and the appropriateness of one's own thinking has to be monitored with attention to perceived cognitive requirements in order to regulate the course of one's thinking in such a manner that costs and risks are minimized.

(b) Human information processing does not work in a one-to-one fashion. There is more than one possibility for coping with a problem. Given the developing repertoire of strategies that a person has available, executive decisions are necessary in order to evaluate the appropriate approach to a problem, to monitor its application and effects and, if necessary, to terminate its application.

(c) Human information processing is fallible and can be improved. This requires the monitoring of one's own thinking, and especially in risky or important situations, a careful regulation of one's problem solving approach. The first empirical results reported by Hesse (13) and Reither (26) with regard to the effects of self-reflection of one's own thinking on problem solving performance indicate a fascinating possibility: the self-guided improvement of our own thinking. This is probably the main merit of metacognition research thus far, that is, having directed the attention of psychological research on cognition towards the self-regulation of human thinking. It is important that human beings understand themselves as agents of their own thinking. Our thinking is not just happening, like a reflex; it is caused by the thinking person, it can be monitored and regulated deliberately, i.e., it is under the control of the thinking person.

If one examines points (a) - (c) with regard to animals, then there is no reason to reserve such a self-regulatory mechanism for the human information processing system. As a consequence one should study the animals' knowledge about problem solving or communication as well as their executive control over such processes.

REFERENCES

(1) Bandura, A. 1977. Self-efficacy: Toward a theory of behavioral change. Psychol. Rev. 84,2: 191-215.

(2) Brown, A. 1978. Knowing when, where, and how to remember: a problem of metacognition. In Advances in Instructional Psychology, ed. R. Glaser. Hillsdale, NJ: Erlbaum.

(3) Butterfield, E.C., and Belmont, J.M. 1975. Assessing and improving the executive functions of mentally retarded people. In Psychological Issues in Mental Retardation, eds. J. Bailer and M. Sternlicht. Chicago: Aldine.

(4) Dörner, D. 1974. Die kognitive Organisation bei Problemlösen. Bern: Huber.

(5) Dörner, D. 1978. Self-reflection and problem solving. In Human and Artificial Intelligence, ed. F. Klix. Berlin: Deutscher Verlag der Wissenschaften.

(6) Dörner, D.; Kreuzig, H.W.; and Stäudel, T. 1978. Lohhausen. DFG-Projekt "Systemdenken." Report 2. University of Giessen, F.R. Germany.

(7) Flavell, J.H. 1971. First discussant's comments: what is memory development the development of? Human Dev. 14: 272-278.

(8) Flavell, J.H. 1979. Metacognition and cognitive monitoring. A new area of cognitive-developmental inquiry. Am. Psychol. 34: 906-911.

(9) Flavell, J.H. 1981. Annahmen zum Begriff Metakognition sowie zur Entwicklung von Metakognition. In Metakognition, Motivation und Problemlösen, eds. F.E. Weinert and R.H. Kluwe. Stuttgart: Kohlhammer, in press.

(10) Flavell, J.H. 1981. Monitoring social-cognitive enterprises: something else that may develop in the area of social cognition. In New Directions in the Study of Social-cognitive Development, eds. J.H. Flavell and L. Ross, in press.

(11) Flavell, J.H.; Friedrichs, A.G.; and Hoyt, J.D. 1970. Developmental changes in memorization processes. Cog. Psychol. 1: 324-340.

(12) Flavell, J.H., and Wellman, H. 1977. Metamemory. In Perspectives on the Development of Memory and Cognition, eds. R.V. Kail and J.W. Hagen. Hillsdale, NJ: Erlbaum.

(13) Hesse, F.W. 1979. Trainingsinduzierte Veränderungen in der heuristischen Struktur und ihr Einfluß auf das Problemlösen. Unpublished dissertation, Technical University of Aachen, F.R. Germany.

(14) Hodos, W., and Campbell, C.B.G. 1969. Scala naturae: Why there is no theory in comparative psychology. Psychol. Rev. 76,4: 337-350.

(15) Kluwe, R. 1979. Wissen und Denken. Stuttgart: Kohlhammer.

(16) Kluwe, R.H. 1980. Metakognition: Komponenten einer Theorie zur Kontrolle und Steuerung eigenen Denkens. Munich: University of Munich.

(17) Kluwe, R.H. 1980. The development of metacognitive processes and performance. Paper presented at the Conference on "The Development of Metacognition, Attribution Styles, and Self-Instruction." University of Heidelberg.

(18) Kreutzer, M.A.; Leonard, C.; and Flavell, J.H. 1975. An interview study of children's knowledge about memory. Mon. Soc. Res. Child Devel. 40 (1, Serial No. 159).

(19) Lüer, G. 1973. Gesetzmäßige Denkabläufe beim Problemlösen. Weinheim: Beltz.

(20) Masur, E.F.; McIntyre, C.W.; and Flavell, J.H. 1973. Developmental changes in apportionment of study time among items in a multitrial free recall task. J. Exp. Child Psychol. 15: 237-246.

(21) Neisser, U. 1967. Cognitive Psychology. New York: Appleton.

(22) Newall, A., and Simon, H.A. 1972. Human problem solving. Englewood Cliffs: Prentice-Hall.

(23) Premack, D., and Woodruff, G. 1978. Chimpanzee problem-solving: a test for comprehension. Science 202,3: 532-535.

(24) Premack, D., and Woodruff, G. 1978. Does the chimpanzee have a theory of mind? Behav. Brain Sci. 4: 515-526.

(25) Pylyshyn, Z.W. 1979. Complexity and the study of artificial and human intelligence. In Philosophical Aspects in Artificial Intelligence, ed. M. Ringle. New York: The Humanities Press.

(26) Reither, F. 1979. Über die Selbstreflexion beim Problemlösen. Unpublished dissertation, University of Giessen, F.R. Germany.

(27) Reitman, W. 1973. Problem solving, comprehension, and memory. In Process Models for Psychology, ed. J.G. Dalenoort. Rotterdam: University Press.

(28) Ryle, G. 1949. The Concept of Mind. London: Hutchinson.

(29) Selz, O. 1913. Über die Gesetze des geordneten Denkverlaufs. Part 1. Stuttgart: Spemann.

(30) Simon, D.P., and Simon, H.A. 1978. Individual differences in solving physics problems. In Children's Thinking: What Develops?, ed. R.S. Siegler. Hillsdale, NJ: Erlbaum.

(31) Sternberg, R. 1979. The nature of mental abilities. Am. Psychol. 34: 214-230.

(32) Thorndyke, P., and Stasz, C. 1980. Individual differences for knowledge acquisition from maps. Cog. Psychol. 12: 137-175.

(33) Ueckert, H. 1980. The cognitive executive: From artificial intelligence toward a phychological theory of consciousness. Paper prepared for the International Congress of Psychology, Leipzig.

(34) Woodruff, G., and Premack, D. 1978. Intentional communication in the chimpanzee: the development of deception. Cognition 7: 333-362.

Animal Mind - Human Mind, ed. D.R. Griffin, pp. 225-250.
Dahlem Konferenzen 1982. Berlin, Heidelberg, New York: Springer-Verlag.

On the Evolution of Cognitive Processes and Performances

F. Klix
Sektion Psychologie, Humboldt-Universität
102 Berlin, German Democratic Republic

INTRODUCTION

Scientific work is mainly painstaking, detailed activity devoted to ascertaining, classifying, and securing facts. Yet from time to time our striving for understanding demands that we try to develop a comprehensive picture of the backgrounds, connections, and general laws relating these facts to one another. The nature of the facts determines how fundamental and objective the laws are which can be developed from them, and the the sciences differ in the basic problems that can be solved on the basis of the factual information they make available. For physics the origin of the universe is a fundamental problem, while the emergence of life on earth is a basic theme in biology. It seems to me that the genesis of intellectual processes and performances, and the origins and principles of development of human intelligence are also major problems for scientific thinking, problems which have a far-reaching theoretical and practical significance. It is to these problems that we shall now turn.

A host of facts about intelligence tests have been collected by psychologists. More than 100 factors of intelligence have been discussed in the literature. Yet have we really come any closer to understanding the functional principles of thinking, which are characteristic of human intelligence, the whence and what for of intellectual effectiveness? Or have we even begun to answer the question of how intellectual creativity is possible and from where it has originated?

We can now ask the following questions: Why have intelligent creatures developed on the blue planet, and what objective laws governed this development? How does a subset of these creatures which call themselves psychologists venture to declare that these laws are explainable, and furthermore to suggest that their own cognitive capabilities will suffice for this purpose? The search for answers cannot hope for success without reliance on fundamental biological laws of evolution, because these laws provide the fundamental source of the capability of learning, which, in turn, forms the basis for the possibility of thinking.

ON EVOLUTIONARY MECHANISMS AND THE ORIGIN OF LEARNING PROCESSES

We can best approach our theme in terms of the origin of life on earth. The biological phase of terrestrial development began, as discussed by Eigen (1), with the interaction between nucleic acids, which are capable of instruction, and proteins. In a molecular sense heredity transmission takes place at this level. At that stage evolution regulates itself according to Darwin's laws. In other words, we are confronted with a breadth of genetic variation on the one hand and selection-based barriers through environmental influences on the other. The composition of the gene pool or a population shifts according to the principles of natural selection. The influence exerted by selective forces changes with the level of evolution. We are especially interested in the point at which these genetic processes lead to instinctive behavior. The components of instinctive behavior regulation have become rather clear thanks especially to ethological research (4,10,14,17). A receptor system (R) divides the flow of stimulation into a hierarchically structured pattern of components which is recognized by an identification mechanism (IM). Subsequently, a behavior program (BP) is activated via a decision mechanism (DM) and is finally activated, also in a hierarchically arranged sequence (Fig. 1). At a higher evolutionary level, an instinct-specific evaluation mechanism (IEM) influences (facilitates or inhibits) the decision for or against a particular unit of behavior. The information utilized in this decision is derived from homeostatic test data. The evaluation mechanism develops, within evolutionary history, into a motivation system. In the case of higher mammals, it is neurobiologically linked with the limbic system. The next question to be asked is then: Why has evolution not ended with the development of instinctive behavior? Or why (in parallel with the development of more complex species) do behavior units dependent on learning come to predominate over those units based on instinct? The answer (at the molar level) is obvious: Genetically wired-in behavior is efficient over long periods only when the environment is also reasonably stationary over equally long periods. Instinctive behavior regulation can be optimally adapted, and in many cases it has proved its mettle and persevered up to the present time. Species capable of primitive learning increase the adaptability of their behavior under changing environmental conditions. At first, sensory identification of external events and motor responses to them are modified by learning. An essential part of this stage is the increasing variability in response to particular stimuli which permit more adaptive behavior. This type of functioning is

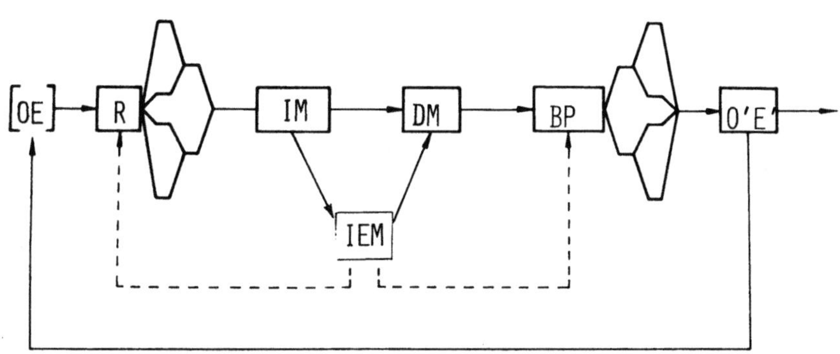

FIG. 1 - Components of instinctive behavior regulation.
OE = object of the environment; O'E' = object altered by action-shifted environment (for other abbreviations, see text).

characteristic of the simplest learning processes of invertebrates, for example, habituation in Aplysia in the experiments of Kandel (5). These animals respond by retracting the gills when the syphon or the mantle is touched. Twenty-four sensory nerves transfer sensory excitation to six motor neurons. The intensity of the response decreases when stimulation is repeated. At the same time the excitatory postsynaptic potential, and the concentration of free Calcium ions, reduce the amount of effective neurotransmitter. An antagonistic process to this is sensitization, as by the production of serotonin which activates synaptic transmission via two intermediate biochemical steps. It seems as though we have here the primitive origins of learning. Sensitization and habituation are fundamental parts of conditioned responses, as investigated by I.P. Pavlov and E.L. Thorndike (11,15). Habituation makes it possible to recognize that stimuli are unimportant, and sensitization permits an increase in the estimated significance of environmental conditions. It is this short-term adaptive function of learning which provides a selective advantage for the animal under changing environmental conditions. This explains why temporary connections between perception and behavior are established. Even the simplest types of learning exemplify this universal principle: learning produces memory traces, and these are the basis of individual memories. In this context there is a sequence of stages in the evolution of learning (16). At first, stimulus identification and motor responses are adaptively connected, while at higher levels the behavior of making decisions becomes differentiated into separate components. Therein lies the basis for the origin of cognitive processes and performances. Parallel to the increasing quantities of stored information, the associative areas of the central nervous system increase. Increasing complexity of animals followed the historical course predicted by Darwin's laws. About 70 million years ago the evolutionary ramification of arboreal insectivores led to the first pongids and hominoids (Proconsul and Ramapithecus) (Fig. 2). As the eyes moved forward on the head, binocular depth perception became possible, and a color sensitive retina also developed. As hominoids changed to an upright posture (during

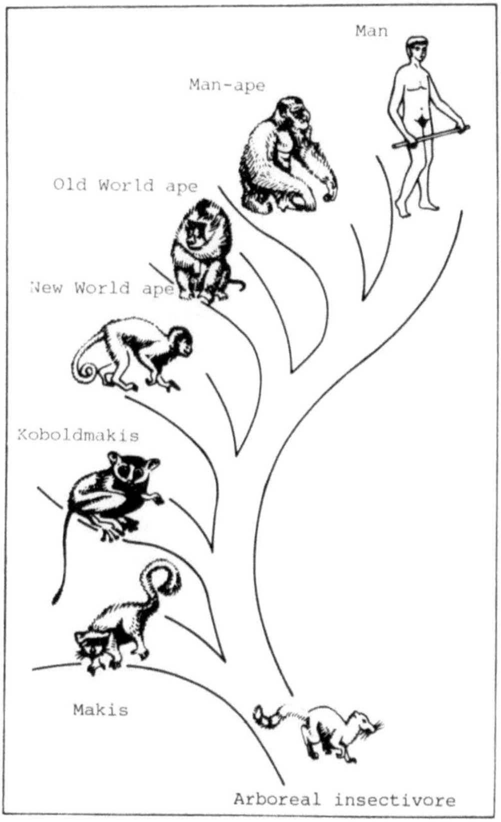

FIG. 2 - Stages in the evolutionary descent of man.

the pre-pleistocene formation of savannahs), the forearm came to be more often in the binocular field of vision. This facilitated precise coordination of eye and hand and was of great adaptive value in the origin of man. At about the same time a precision grip appeared, which is common to pongids and human beings, and this provided an important precondition for tool making.

Without being able to discuss all the intermediate stages, I have touched on a few of the functional principles governing the evolutionary transition from animal to man. Let us now turn to some of the characteristic capabilities of creatures living at the beginning and at the end of that period of time, which extended from about 3.5 million to roughly 250,000 - 300,000 B.C.

THE ORIGIN OF MENTAL PERFORMANCE AND LINGUISTIC COMMUNICATION BETWEEN THE BEGINNING AND THE END OF THE TRANSITION PERIOD FROM ANIMAL TO MAN

Let us assume that living pongids have cognitive constitutions which are comparable to those present at the beginning of the transitional period from animal to man. The following features call for our attention: The two most important preconditions for the origin of language existed, but there was no language in the proper sense.

The first precondition was the capability of acquiring concepts. Köhler, Ladygina-Kohls, and Rensch et al. (7,8,13) showed that chimpanzees are able to form concepts, such as the concept of a screwdriver. A number of individually very different objects have in common a few characteristics. They thus have the same value and usefulness for a certain operation, which is the essence of a simple concept (Fig. 3). It is the most central element of thinking and was in existence before the appearance of language.

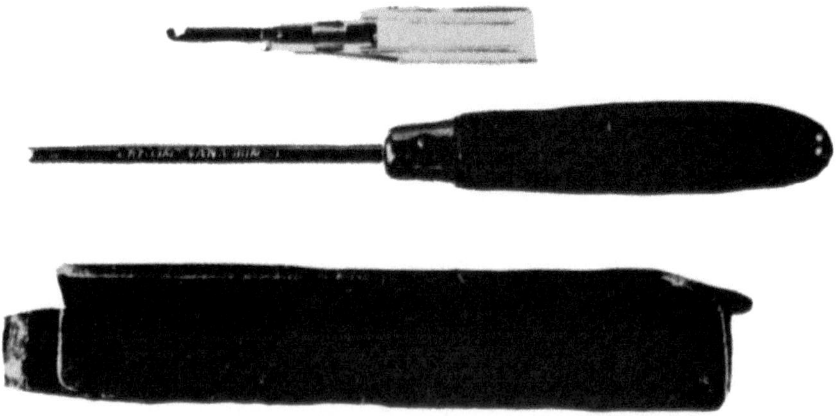

FIG. 3 - Three different types of screwdrivers.

The second precondition is the emergence of a communication system based on signals which in turn depend on learning. These are unequivocally in existence. According to Lawick-Goodall (9), apes use a number of signals which serve for communication with conspecifics by means of stylized patterns of behavior. Such communication signals arise as inherited behavior patterns change their function as a result of learning.

Yet these two preconditions remain separate functions. The communicative signals serve merely to influence immediate behavior. They are learned in social relationships. Concepts, that is classifications of objects, originate in action, in activity. Although Premack and Fouts et al. (3,12) have demonstrated that chimpanzees can learn symbols for concepts, they do not do so in the wild. Communication and cognition remain separate activities. It was of great importance in human evolution when these two began to be used together, and the question naturally arises: How did this important interaction begin? It is a natural suggestion that coordination of group activities and social relation was crucial. Or to put the matter in different terms: The beginning of a division of labor provided the decisive impulse for the development of language through the integration of concept and signaling gesture. F. Engels (2) wrote: "Labor first, then and with it: language." The results of the division of labor in its early stages are provided by the record of Celtic cultures, in the period from the first appearance of pebble-tools (Fig. 4) through working implements constructed for particular functions up to the constructively manufactured saw. Almost two million years elapsed during the development of this type of social division of labor. This anthropological progression occurred between the time of skull 1470 to the appearance of Homo sapiens.

It is necessary to assume that in all manufacture of tools a plan of action was present in the cognitive background of tool construction. In the most recent example (the stone-saw), however, a special type of manipulation was required: the material

FIG. 4 - Pebble-tools of Homo erectus (ca. 2 million years BC).

was first broken apart into pieces, well-defined parts were constructed, and these were then recombined for a new purpose (Fig. 5). (An analogy springs to mind in the manner in which individual elements are combined to produce communication signals.) The hierarchical organization of action plans is also unmistakable. The GDR excavations in Bilzingslaben suggest that even in the time of Homo erectus individual specializations of function existed in the manufacture of tools, and this was definitely the case for the Cromagnon people. We should recognize that this era included both glacial and interglacial periods. Hunger, cold, and misery made the highest demands on motivation to achieve, doubtless exhausting and restimulating it again and again. Successful hunters, gatherers, trappers, and medicine men were required. The capabilities of the group, the tribe, or the clan are differentiated in accordance with economic, social, or cultural needs. Are cultural needs truly important? Yes, indeed. These needs are based upon the strongest of socially conditioned motivation patterns. They are intended to protect the community, to cure the sick, to influence the future,

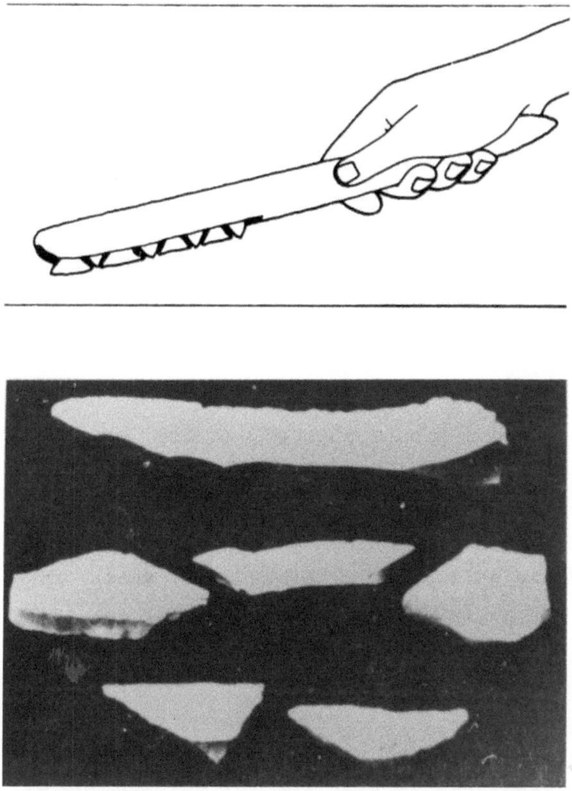

FIG. 5 - Stone-saw and material used for its construction (30 thousand years BC).

and to stabilize the community's pattern of life. The totum and taboo mark the first rules of social coexistence.

Ritual acts are an essential element of archaic thinking. We shall look at some of these qualities, because the origins of modern scientific thinking are rooted in them.

SOME MENTAL AND COGNITIVE TRAITS OF ARCHAIC THINKING

In the history of mankind two major strategies have been developed to meet what is always felt to be an urgent need to control future events. One of these needs consists in influencing the future by means of witchcraft and magic expressed in ritual behavior in order that hoped for events can be brought closer to reality and feared events can be banned. Attempts are made to gain mastery over the forces of nature by thought and by symbols expressing thoughts. The second strategy is a sort of reversal of the first. It consists of recording consistently repeated events that are regular natural occurrences in order to be able to get a grip on them and comprehend the fixed invariant and predictable features of the world. This is why registrations were created to show the cyclic nature of the tides, phases of the moon, bird migrations, or when the star Sirius rises above Memphis every 365 days thus predicting the floods of the Nile.

Both these strategies have the same purpose, namely, to cope with emergencies and, above all, to reduce uncertainties about the future. To eliminate uncertainties, however, means to obtain information for the sake of deciding upon appropriate future actions. And the gaining of information by means of cognitive processes is the essence of human thinking. The more effectively this takes place, the higher the level of intelligence.

The results of the second strategy have become historically evident (6). One was a principal impetus for the development of writing and the alphabet, the other for the development of the concept and system of numbering.

These two processes operating harmoniously together in writing and calculation are as much the result as the cause of all intelligent performances that are specifically human. Now I would like to demonstrate that these two major intellectual achievements of mankind, writing and numbering systems, have emerged from the same kind of cognitive functions.

The first achievement, the origin of written language, developed in several stages. One preliminary stage is represented by the vital imagery of cave drawings which depict mainly scenes and occurrences expressing emotional thinking together with vital strength and specific types of information. The evolution of writing leads, at the first stage, to a certain neglect of the emotional and a certain emphasis on the informational content of events. The presentation of a tribal feud by a Sioux indian, as well as the disk of Hephaistos found in Crete (Fig. 6),

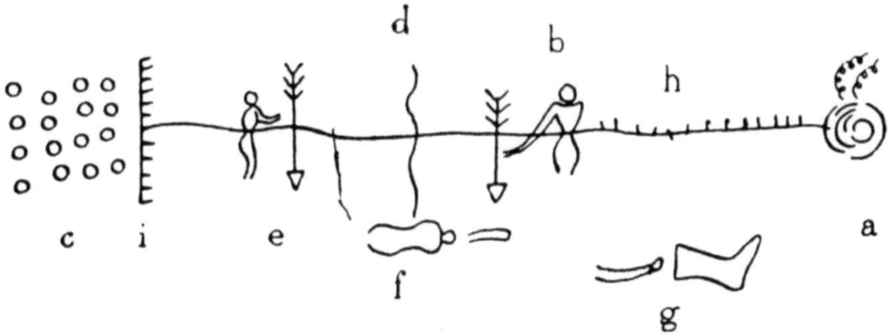

FIG. 6 - Presentation of a tribal feud by a Sioux indian.

(originating from the culture of Minos and not yet deciphered), show quite analogous forms of expression. In both, scenes are partitioned into stylized pictures. Their composition, or concatenation, constitutes the text. The capability of expressing thoughts in this manner requires a considerable plasticity of thinking and communication.

At the next stage we no longer find only scenic symbols for a given event, but also a symbol for the concept. The six lotus flowers on the ornamental tablet of King Namer (2800 BC) designate six times the concept of 1,000 (Fig. 7). And the ancient Chinese choice of symbols for concepts makes us recognize the same principle at work. At this stage, too, the expressive power of writing is limited. If there is a single symbol for each concept, how can concatenations between them be expressed, or how can their relation to thoughts about previous or future occurrences be conveyed? A new problem thus arises: A system

FIG. 7 - Ornamental tablet of King Namer (2800 BC).

of symbols is needed which is capable of expressing, with a limited effort, a potentially infinite variety of mental experiences. The ancient Egyptians discovered the first way to solve this problem. It consisted in transferring one already known solution to another case, to an analogous level. Spoken language was available. Its grammar was capable of expressing any thoughts contained in the people's conceptual dictionary. Therefore symbols for the sounds of speech and not for concepts were the earliest to appear in writing. This is the Rebus principle, where, for example, in German a watch (Uhr) and a leaf (Laub) are drawn in order to express the concept of holiday (Urlaub). Although the expressiveness of written language was thus raised to the level of the spoken language, this procedure was unwieldy. It required a huge effort to learn thousands of symbols. What was still lacking was the inspiration for an ingenious solution that would achieve simplicity.

This final step was taken three times within a historically short period on the eastern border of the Mediterranean Sea between 1500 and 800 BC. The fact that this occurred in urban settlements such as Byblos, Tyros, Ugarit, and Sarepta suggests that it was stimulated by a common socioeconomic condition. It was the discovery of an alphabet of the 22 letters standing for consonants introduced by the Phoenicians which provided the ancestral alphabet for the Greek, Cyrillic, Etruscan, Celtic, Latin, and Gothic letters. Again the cognitive procedure underlying this development was the subdivision of complex symbols into simple elements and their recombination into new forms. This makes possible the production of any kind of word which can be spoken. The principal influence in this final stage arose in the sumeric, Assyric-Babylonian culture. At the second stage we can see how basically similar cognitive processes (namely, subdividing complexes by abstraction, combinatory composition, concatenation, and hierarchical order) lead to a second system of symbols which demonstrates one of the major reasons for the increasing capabilities of human intelligence. I am referring to numbers and number systems.

NUMBERS AND NUMBER SYSTEMS AS EXPRESSIONS OF COGNITIVE DISPOSITIONS

A precondition for the comprehension of numbers is the capability of differentiating between quantities. From the standpoint of evolutionary theory, this arises in connection with perceptual capabilities found in invertebrates and birds, for example, in bees and pigeons. The actual counting requires the isolation of the characteristics of the thing itself from the property of quantity. In the formation of writing this is called sensory isolation (as a type of abstraction), and we are already familiar with the same procedure from the formation of concepts and the emergence of communicative signals. The same process is exemplified by the cognitive subdivision of the impression of a volume when a number of examples are compared. The essential cognitive difficulty experienced in counting arises from the individually different appearances of the things to be counted. Similarly, as with writing, in the development of numbering systems, a shift occurred in designating quantities, namely, via the construction of auxiliary sets. This development makes objective, so-to-speak, the outcome of the sensory isolation by making it visible. When the Wedda of Ceylon count nuts, they attach a stick to each one. The question: "How much have you got?" is answered by "So much" while the sticks are displayed. Thirty to forty thousand years ago counting was probably also done in this way. As evidence we may turn to the carvings on the legbone of a young wolf (Fig. 8) which lived some 40,000 years ago in present-day Czechoslovakia. Certainly at that time numerals did not exist. The names of numerals presumably arose in a fashion analogous to the origin of words in writing. How many words for figures are required? Must there be a new one for every countable quantity? That would yield an infinite quantity of number words which would be impossible to learn.

The sensory isolation made countability possible. The concatenation of primary numerals into groups and then into new groups of groups was the next step which was first taken by the ancient Egyptians (Fig. 9). First we find a mixture of pure symbols

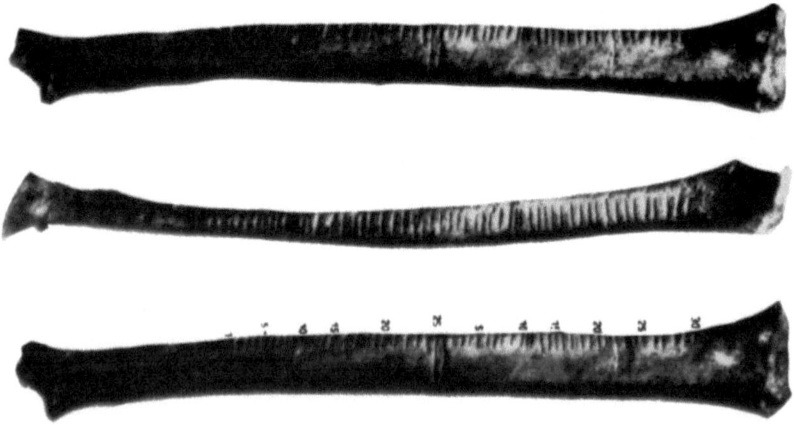

FIG. 8 - Counting tool made from legbone of a young wolf (ca. 40,000 years ago).

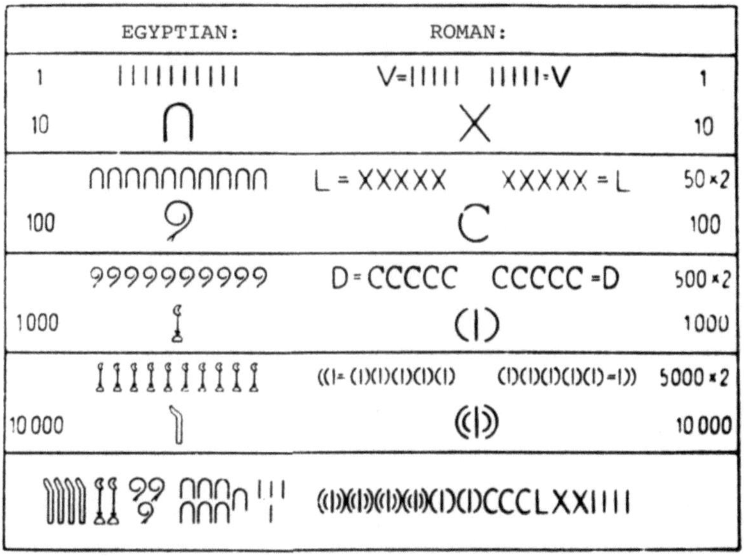

FIG. 9 - Egyptian and Roman principles of number groupings.

and pictograms as were later elaborated by the Mayas of Central America. We see the same in Chinese pictograms. The amount of concatenation differs greatly among these peoples, but the principle of grouping countable units is invariably present. The numbers of fingers or toes may have been a primitive basis for counting: In the languages of Minos and in ancient Greek we find groupings of five.

The Egyptians and the Romans used the number 10. The Copok in Russian is based upon the serko (that is, the animal skin). Animal skins were traded in bunches of twenty. Thus even today the numerals of various languages demonstrate old groupings used in counting (in Scottish and Irish 20 = fiche and 80 cithre fiche, in French quatre vingt means four times twenty). The function of the bunching or concatenation even of new units is obvious: The quantity of numerals to be written diminishes and the learning effort decreases while an equal cognitive efficiency is maintained. The hierarchical grouping of cipher concepts (Fig. 10) made an infinite variety of quantities cognitively manageable. Human intelligence uses these strategies of simplification to cope with otherwise unmanageable cognitive demands.

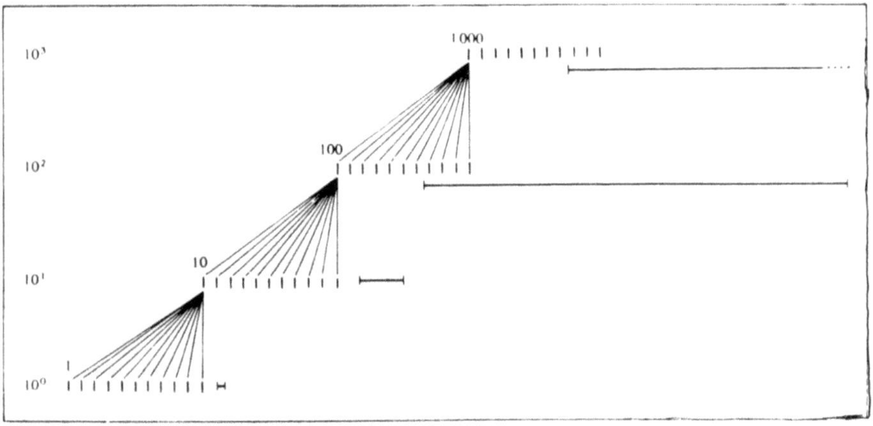

FIG. 10 - The hierarchical grouping of cipher concepts.

This is demonstrated by the process leading to the highest step to date in the development of counting systems. This was first taken by the Sumerians who had a sexagesimal number system with sixty as the unit. Figure 11 shows their figures. What they achieved was an immense simplification of presenting ciphers which became possible through the interplay of counting and writing (Fig. 11, bottom). On the left we recognize the (old) mass presentation of the figure 410,482. On the right is the same figure presented in a much simpler fashion. All modern numerical systems use this general method of simplification.

Now the question arises: Why has the cognitively simplest way of presenting numbers prevailed in the course of history? Only because it is more simple and cognitively more easily manageable? I believe so, but one must explore this reason in greater depth. The simpler cognitive representation of a problem leads to the possibility of solving it with the requirement of fewer operations. And, the simpler the representation, the

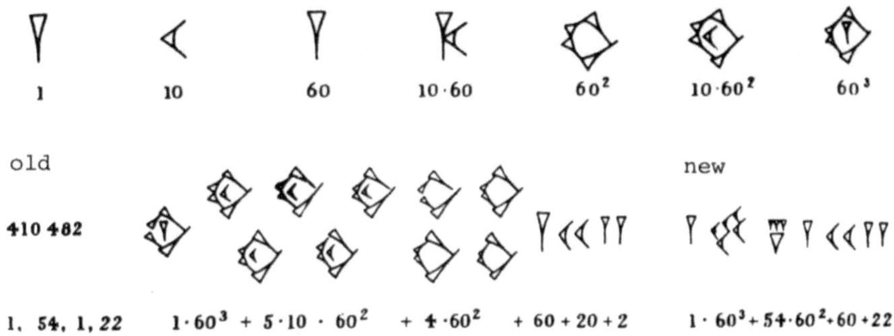

FIG. 11 - Sumerian sexagesimal number system (see text).

greater is the degree of difficulty that can be managed with the same effort. This is the functional principle by which it has been possible to achieve an increase in intellectual effectiveness, that is, in cognitive power over the course of human history. By taking steps toward more effective number systems, increasingly difficult problems became mathematically manageable. The same principle may well operate in gifted persons. One such example was the discovery of the sum formula for arithmetic series by the sixteen year old C.F. Gauss. Gauss found that the simplified representation of the problem required fewer operational steps and therefore he arrived more quickly at his goal and overcame greater degrees of mathematical difficulty than contemporaries of his age. A few years ago our institute carried out problem solving experiments with highly gifted mathematic students and compared them with normal students. We arrived at the same conclusion: The more gifted use simpler methods of representing problems and these required less effort in reaching a solution. This suggests a possibly fruitful approach to the principles of measuring intelligence.

The Sumarian number system functioned, in all probability, in a manner similar to the cognitive strategies of highly gifted persons. Here are a few examples to support this hypothesis: Exchange, trade, and fares are based on division of objects into categories. The most essential measuring system in astronomy is the division of time. The unit of 60 introduced by the Sumarians has the property that it can easily be related to important natural events. The circle has 6 x 60 degrees, the minute 60 seconds, the hour 60 minutes, twice 1/10 of 60 is the duration of the tides (tides = ebbs and flows = 6 hours each), 1/5 of 60 is the number of hours in a day, 1/2 of 60 is the number of days in a month, and 6 x 60 days constitute one year. This provides the greatest simplicity in measuring space and time. Even Ptolemy counted according to the sexagesimal system. By means of their reciprocal tables these Sumarians reduced division to multiplication. The logarithms and the Leplace-transformation provide the same type of simplification. Trying to carry out multiplication with Roman numbers allows one to experience how representation of a problem has a very great effect on its solubility.

A final step in the development of cognitive capabilities remains to be explained.

THE ORIGINS OF META-LEVELS OF HUMAN THINKING AND THE DISCOVERY OF LOGICAL FORMS

Writing and the number system represent the first great advance in purely cognitive processes. Writing and numbering are the objective results of cognitive processes. They are ways of expressing social consciousness. These graphic results of intellectual processes became objects of human perception just as did natural phenomena many centuries earlier. It is not surprising that human cognitive activity would begin to explore its own procedure just as it had previously explored the laws of nature. Cognitive processes tried to understand its own method of functioning. This sort of activity by human intelligence leads to meta-levels of thinking, to the discovery of logical forms and of deductive conclusions. When viewed historically, this step in intellectual development is inseparably linked to Greek history, for it was in Greece that the transition occurred from archaic thinking to scientific, and especially to natural scientific thinking. It is in a way surprising that the basic logical forms of human thinking were not found from dealing with numbers, not through mathematical thinking, but from language. Plato recognized as reflected in the Socratic dialogues that behind the words lie the actual deciding influences in thinking, namely, concepts. The epoch making analysis of concepts was found by Plato's Macedonian student and Alexander's tutor and fanatical Hellenist, Aristotle. In his "Topic" and in "Hermenaia" he wrote that cognition involves logical structures which are essential in human thinking and which must be recognized as equal in importance to the laws of nature. Both of these are bound to follow just as a train follows the rails upon which it has been placed. The syllogistic rules, the Modus Barbara, such as the Modus ponens, determine such consequences of our thinking. These laws of thinking eliminate any doubt about basic logic.

The gateway to the future of scientific thinking was opened by Aristotle through his analysis of the basic forms of logical statements. Through an understanding of the role of free variables in a group of arguments he discovered a principle for the formalizing of linguistic statements. Today computer linguistics have proved the significance of such thoughts in connection with natural language processing through computers. The laws of nature have been combined with the laws of human thinking as a theme for scientific research. What was it that F. Engels (2) wrote in 1880? "And thus it does not surprise us that the laws of thinking and of nature must necessarily coincide as soon as they are correctly recognized."

In the remote past the development of tools by the Celts helped to increase the productivity of human labor; nowadays this is accomplished as a result of the development of a high order of intelligence and reliance on computer programs. Human intelligence is now ready to use its own effective procedures as a means of exploring the methods by which the human brain actually operates. This is an important task for the future of psychology, and insofar as it succeeds psychology will find a new place for itself among the sciences by making a major contribution towards solving this basic social problem.

This could be the end of this paper, but the most difficult of all issues concerning human intelligence has not yet been discussed, namely, how can new achievements originate through the process of thinking when no prototypes exist (for example, the insights of Copernicus or Leibniz which are so original that they race ahead of general human knowledge by many generations)? Let us pose this final question as Emmanuel Kant would have posed it: How is creativity possible in human thinking?

ON DEVELOPMENTAL CONDITIONS FOR CREATIVE ACHIEVEMENTS IN THINKING

I have explained how intelligence expresses itself at a high level of development in the quality of thinking. In this context this quality is essentially determined by the cognitive representation of a problem, which in turn determines the simplicity of finding a solution. The simpler the representation, the higher the performance of intelligence, and the more extensively can problems be handled with the same effort. Intelligence manifests itself in the methods of producing results. What we call representation is, in this context, a result of processing information in relation to a problem situation. What happens, however, if different forms of representation occur together in a problem solving situation? How is it possible to find the simplest solution? I would like to demonstrate how this is possible by means of a few examples which illustrate the source of creative accomplishments in thinking.

For reasons which are connected with archaic thinking in our prehistory, we can think both in a vivid-pictorial or in a logical, conceptual form of representation. A circle has its perceptive and its conceptual form of representation which differ greatly (Fig. 12). This seems trivial, but the change between

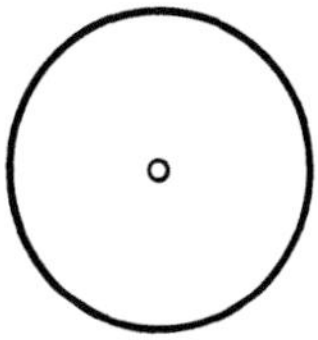

$$x^2 + y^2 = R^2$$

Equation of a circle
Cartesian coordinates

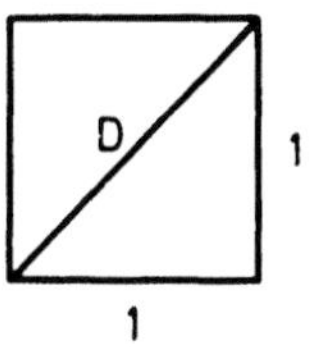

$$1^2 + 1^2 = D^2$$
$$D = \sqrt{2}$$

FIG. 12 - Pictorial and logical-conceptual representations of a circle and a square.

these two kinds of representation had led to the greatest discoveries in the history of mathematics. Here is one example: Euclid was very well aware of the line relations of squares and triangles. By applying, along with others, the Pythagorean theorem he was able to demonstrate that the diagonal of a square cannot be expressed as a simple function of the length of one side (with numbers known to the Greeks). But a solution to this problem is now possible by means of irrational numbers which were unknown to the Greeks. The same principle of the alternate internal representation of cognitive structures is illustrated by the use of pictorial or conceptual forms of representation in creative performances. Ingenuity often arises from mapping knowledge of one problem onto another problem. Such mapping gives rise to the process of thinking via analogies and making inferences about new problems on the basis of solutions to similar problems. For instance, it has been reported that Thales intended to measure the height of an Egyptian pyramid. Figure 13 demonstrates his method of procedure. It is just as simple as it is ingenious, and it is ingenious because Thales reduced the seemingly insoluble problem into a simple childlike one. He waited until the appropriate time

a

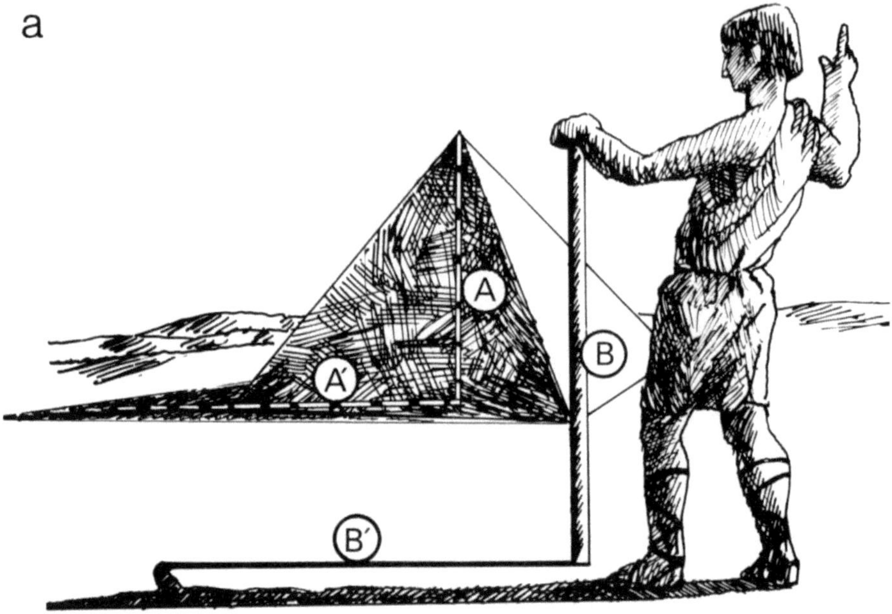

A = height of pyramid, A' = length of pyramid's shadow
B = height of stick, B' = length of stick's shadow

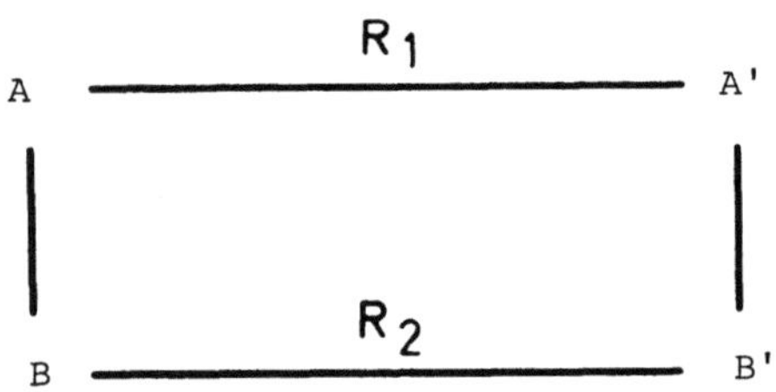

FIG. 13 - Depiction of Thales' measurement of the height of a pyramid where a) A:A' = B:B', or b) $AR_1A' = BR_2B'$.

when the length of his stick equalled the length of its shadow. At the same time the length of the pyramid's shadow had to equal its height. The cognitive basis for this solution is obvious, a relationship between A and A' is mapped onto a relation of B to B'. This realization of the identity $R_1 = R_2$ is in fact a most ingenious solution. What is most important about it is the great simplicity of the cognitive functions which produce a creative achievement of this kind.

An accomplishment of a completely different sort is based on similar lines of thinking. As a protagonist of the Copernican world picture, Galileo had to argue against the objections raised by his Aristotelian contemporaries. They argued that in the case of the rotation of the earth, a vertically falling stone hits the ground somewhere other than directly below the point from which it was dropped. Galileo refuted that theory by the example of the moving sailboat (Fig. 14), showing that this result does not apply because the boat is moving. Mapped onto the level of knowledge of the earth's rotation his conclusion was that the objection did not apply. Yet the relations R_1 and R_2 are identical.

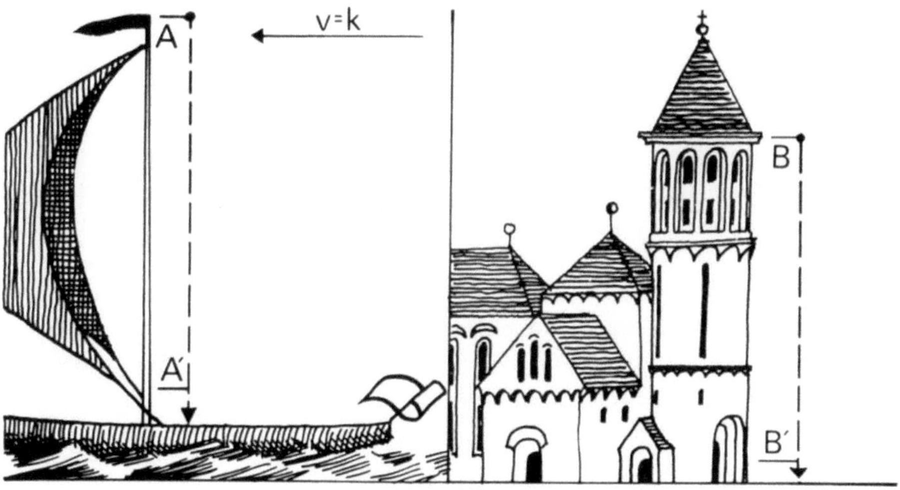

$$\frac{\begin{array}{c} A\,R_1\,A' = B\,R_2\,B' \\ A = A' \\ B = B' \end{array}}{R_1 = R_2}$$

FIG. 14 - Analogous conclusion of a "Gedankenexperiment" by Galileo (see text).

Again we are faced with the same structure of an analogous conclusion as the basis for a creative achievement. A great many similar examples taken from the history of science also demonstrate this point: e.g., when Archimedes calculated a parabolic segment which was not possible by available, pure mathematical means, he arrived at a solution by mapping the whole onto the laws of the lever; or when Newton and Leibniz derived differential quotients quite independently of one another using different kinds of representations. All these examples indicate that there are mapping procedures, within and between cognitive representations of problems. The mapping between logical and pictorial (or iconic) representations is a special example of such procedures, which have a good deal to do with the generation of analogies. The examples indicate that this is a source of ingenious or creative thinking. And who would not recall at this point the famous analogies involving trains and elevators contained in Einstein's thought experiments?

It is therefore not true that intelligence and creativeness have little in common, nor is it the longer memory span or differentially structured or more complicated mental operations which constitute the basis for our ability to arrive at ingenious intellectual achievements. Whatever we discover about the historical development of our mental processes can also be applied to individual cases and individual talents. The representation of problems is one element of working memory, along with information processing activities. These structures determine the effort required to find solutions. Their simplicity and usefulness reflect the degree of intellectual power brought to a particular problem.

In the course of our extensive excursions we have, again and again, become acquainted with similar principles of cognitive function in different mental spheres, different times, and different regions. These involve sensory isolation, combinations and concatenations of individual elements, abstract compression, and simplifying of operations by means of hierarchical organization. All this demonstrates how within the huge wealth of variety in mental life, basic laws of cognitive processing remain invariant just as does the chemical composition of living organisms. Biological systematists had already performed their major work before biochemists began clarifying the elementary mechanisms found in living organisms. The same basic process has now begun in psychology through the microanalysis of cognitive processes. These extend, as our Dahlem Conference demonstrates, up to the elementary analyses of creative thinking. I believe that in our field of psychology we are about to gain insight into the fundamental mechanisms which control intellectual life. The combinatorial diversity of cognitive elements in thinking is analogous to the role of combinations of amino acid sequences in producing the diversities of bodily structures and functions in living organisms. For this reason we should not be impatient when psychologists use small puzzles and plays on words in their analyses of human thinking. The laws of heredity were, after all, worked out not by studying dairy cows or fattened pigs, but by quantitative experiments with small black flies, intestinal bacteria, or phages. The practical results from biological research are widely known and recognized, and if we are not altogether mistaken, the results of research on the fundamental laws of intellectual processing will be no less important in future years. We have just cast a glance back through time at a huge natural experiment which produced human society. In the course of that experiment there emerged a phenomenon which we call thinking, a process by which Homo sapiens arrived at self-consciousness.

Acknowledgement. I would like to thank D.R. Griffin for his helpful rewording of the final version of this paper.

REFERENCES

(1) Eigen, M. 1972. Molekulare Selbstorganisation und Evolution. Informatik. Nova Acta Leopoldina 37/1 (206).

(2) Engels, F. 1952. Der Anteil der Arbeit an der Menschwerdung des Affens. Berlin: Dietz-Verlag.

(3) Fouts, R.J. 1974. Language: Origins, Definitions and Chimpanzees. J. Human Evol. 3: 350-358.

(4) v. Frisch, K. 1953. Aus dem Leben der Bienen. Berlin: Springer-Verlag.

(5) Kandel, E. 1979. Small systems of neurons. Sci. Am. 241 (3): 60-78.

(6) Klix, F. 1980. Erwachendes Denken. Berlin, GDR: Deutscher Verlag der Wissenschaften.

(7) Köhler, W. 1917. Intelligenzprüfungen an Anthropoiden. Berlin: Springer-Verlag.

(8) Ladygina-Kohls, N.N. 1958. Die psychische Entwicklung im Prozeß der Evolution der Organismen. Moscow: MIR.

(9) Lawick-Goodall, J. 1975. The behaviour of the chimpanzee. In Kommunikation und Verhalten, eds. G. Kurth and F. Evol-Eiberfeldt. Stuttgart: Gustav Fischer.

(10) Lorenz, K. 1957. Methoden der Verhaltensforschung. In Handbuch der Zoologie, eds. G.H. Helmcke et al. Berlin: Springer-Verlag.

(11) Pavlov, I.P. 1953. Gesammelte Werke, vol. 3. Berlin: Akademie-Verlag.

(12) Premack, D. 1976. Intelligence in Ape and Man. Hillsdale, NJ: Erlbaum.

(13) Rensch, B. 1968. Handgebrauch und Verständigung bei Affen und Frühmenschen. Bern: Huber.

(14) Tembrock, G. 1972. Tierpsychologie. Wittenberg: A. Ziemsen.

(15) Thorndike, E.L. 1913. The Original Nature of Man. New York: Wiley.

(16) Thorpe, W.H. 1956. Learning and Instinct of Animals. London: Methuen.

(17) Tinbergen, N. 1952. Instinktlehre. Berlin: Verlag Parey Hamburg.

Animal Mind - Human Mind, ed. D.R. Griffin, pp. 251-268.
Dahlem Konferenzen 1982. Berlin, Heidelberg, New York: Springer-Verlag.

Study of Vertebrate Communication – Its Cognitive Implications

C. G. Beer
Institute of Animal Behavior, Rutgers University
Newark, NJ 07102, USA

Abstract. Study of communication in animals involves two directions of cognitive implication: cognitive assumptions of the investigator affect how social interaction is perceived and described, hence the selection of fact, formulation of question, cast of explanation; accounts of communication behavior may premise inference to cognitive capacities in the animals observed. Examples of the latter include how an animal's sensory capacities and perceptual selectivities imply bounds to its experience and knowledge; how memory is reflected in the statics and dynamics of social interaction; how observations of the use of signals have been interpreted as support for various conceptions of animal communication, which include expression of emotion, of behavioral tendency, and of semantic sense. Some examples imply intention; others involve aesthetic considerations. General conclusions are that vertebrate communication comprises more diversity in form, function, and underlying cognitive correlates than has been traditionally supposed; and that the kind of mentality inferable from communication behavior of animals compares to the human roughly in accord with phylogenetic affinity and the degree of similarity in social life.

INTRODUCTION

"... the subtlety and strength of consciousness are always in proportion to the capacity for communication of a man (or an animal), the capacity for communication in its turn being in proportion to the necessity for communication..." (Nietzsche, quoted in (7)). What kind of statement is this? Is it a generalization from facts? If so, what are the facts? In 1882,

when Nietzsche made the statement, there were few precise comparative observations of animal communication, and even today one would be in a quandary if required to give estimates of the "subtlety and strength of consciousness" in various creatures. Could the statement be a hypothesis then? If so we can ask what kinds of fact would test it, and again be in want of an encouraging answer.

A third possibility is that the statement expresses what Dennett (7) calls a "stance," a way of regarding the subject of animal consciousness and the phenomena of animal communication, affecting which concepts and conceptions can be brought to bear, what perception and description can single out for notice, hence, what can count as a fact and an explanation. Every study starts from a stance of some sort.

I therefore take my title to refer in two directions. On the one hand there are the implications for the study of communication that derive from constraints set by the cognitive capacities and biases of the investigator. We can come to know only what we have a mind to know. On the other hand there are the implications in the hope of which this conference was convened: inferences from observations of animal communication to conclusions about animal minds. Such inference could well cause one to change one's conception of animal communication, to adopt a new stance toward what one sees taking place in social interaction, or to see the behavior differently from the previous point of view. But a stance of some sort must always be prior to the seeing. Therefore I shall begin with some comment about cognitive implications of "the beholder's share" in the study of communication. Then I shall look at some examples of vertebrate communication, selected (by me if not by nature) as possible "windows" on animal mentality (10).

FRAMES OF REFERENCE

"Isn't it funny how everyone else speaks with an accent?" (This from a New Zealander recently returned home from abroad).

Similarly the scientist who sees how the views of others are distorted by assumption and preconception may be blind to the beam in his own eye. The confidence with which we pursue present studies matches the confidence with which we dismiss the authors of past attempts. For example, Gall, the founder of phrenology, has claim to be "the writer of the first systematic animal psychology," yet no ethologist bothers reading him now; the faculty psychology which was his frame of reference is no longer regarded as viable. Similarly Romanes provides illustration of the errors of anthropomorphism and anecdotalism, rather than truths reached by his "ejective" approach to animal mentality. Behaviorism, of course, put everything right by banning mentalistic talk; yet one can still argue that behaviorism was really the old associationistic psychology got up in modern dress, that its empiricist and environmentalist assumptions were inherited doctrine rather than inductive finding, that the laws of learning were the principles of associationism with stimulus and response in the places of sensation and idea.

Watson's redefinition of psychology as "the science of behavior" was anticipated by McDougall. However, as in so much else, the two disagreed about what should count as behavior: for Watson it was movements of the body; for McDougall it was action pursued with purpose. This distinction between behavior regarded as movement and behavior regarded as action continues to be a controversial issue, especially in philosophy, where it has accumulated a considerable literature of relevance to our subject (e.g., (13)).

This question of the individuation of behavior bears on how we describe social interaction and hence what our descriptions of it imply. In a recent attempt to "provide a new basis for the quantitative study of overt behaviour on a level of objectivity and precision which is equal to that of other biological sciences," Schleidt and Crawley (19) argue a case for keeping "the descriptive process exclusively within the domain of form." Consider how this approach might apply to some of the problems in communication study, including that of human speech and writing. Presumably it would dictate that, in Peirce's jargon, a type be defined in terms of the characteristics of its tokens; yet Peirce himself argued that something could be a sign only by being interpreted as such, from which it should follow that "... the notion of a type will have to be taken as primitive, and that of a token defined in terms of it" (2). Applied to analysis of spoken or written language the approach would presumably take the "surface structure" and wring Markov chains out of it, a procedure which Chomsky showed to be incapable of arriving at the grammatical rules of the language. Even so, Chomsky was in harmony with the structural approach in so far as he argued the independence and primacy of syntax in relationship to semantics. On this issue his theories have met resistance from several quarters. For example, in a discussion of "artificial intelligence as philosophy and psychology," Dennett (7) commented: "... it is notoriously hard to see how Chomsky's early syntax-driven system could interact with semantical components to produce or comprehend purposeful speech." Part of the problem here is that questions tend to get confused with one another. Chomsky (5) treats the ontogenetic acquisition of language, the evolutionary derivation of language, and the problem of representing the general characteristics of language as dependent on an abstract system of generative grammar, as if these were all one and the same kind of question, which they are not.

"What counts as a fact depends on the concepts you use, on the questions you ask" (17). But the questions you ask depend upon the descriptions you start with. If in describing communication behavior you use action terms implying intention, you may pose questions about what the animals were trying to do, which would not be the case were the description to refer solely to changes of position of the body. Compare "he raised his hand" and "his hand rose" - Hornsby (13) argues that the former implies an anterior trying which the latter does not. Now we commonly say of an animal something like "it lifted its tail" rather than "its tail went up." Is the former merely a careless expression which should be read or heard as the latter? Or do we, at least on some of the occasions when we describe behavior as action, intend to ascribe the intention normally implied in such description?

Dennett (7) argues that the "intentional stance" can be usefully taken even towards machines: "The inescapable and interesting fact is that for the best chess-playing computers today, intentional explanation and prediction of their behavior is not only common, but works when no other sort of prediction of their behavior is manageable." The consequence of talking as though a machine can have intentions, desires, beliefs, and reasons should not, Dennett says, get us hung up on the question of whether such could really be the case: the mentalistic talk is a pragmatic strategy justified to the extent that it enables one to master the machine's moves. To take this position with regard to animal communication no doubt gives less than a cognitive ethologist might hope from inference premised on overt behavior. But at least it opens a way to the use of cognitive concepts in discussion of animal behavior and may be the best such means to be had.

Both "communication" and "cognition" offer considerable latitude for the choosing of initial positions. Most general discussions of animal communication begin with the question of definition and admit to offering a less than definitive answer. Few would disagree that animal communication has to do with characteristics and behavior " designed" for signalling; but opinions differ about what is signalled and to what ends, whether motivational connotations apply and how, who benefits and at whose cost, and other questions. Cognition covers the capacities of perception, memory, and thinking, but with emphasis differing from one writer to another. Perhaps the majority give pride of place to thinking and distinguish it categorically from sensation and perception. In contrast McDougall made perception central to his concept of cognition (knowing), which together with conation (willing) and affection (feeling) constituted his notion of instinct: "... an inherited or innate psychophysical disposition which determines its possessor to perceive, and to pay attention to, objects of a certain class, to experience an emotional excitement of a particular quality upon perceiving such an object, and to act in regard to it in a particular manner, or at least to experience an impulse to such action." But even the sharp separation of knowing from emotion and motivation has its critics, who think that human preoccupation with intellectual abstraction and the written word hinders appreciation of how what we know is interfused with how we feel and what we want. So Langer (15) makes feeling "the mark of mentality"; she echoed Brentano, who made intentionality the mark of the mental; and Midgely (17) argues that reference to consciousness and motive must attach to most description of behavior if it is to be intelligible, especially in the case of communication.

Midgely's reasoning is relevant to a current issue about whether communication behavior is for conveying information (22) or for manipulation (6). Dawkins and Krebs argue from the "selfish gene" viewpoint of natural selection; hence their use of the term "manipulation" is metaphorical rather than literal, as is implied by their statement: "Needless to say, there is no implication of conscious prediction." But, as Midgely insists, there is the question of motivation to which the notion of manipulation might apply in its usual sense, which does imply intent. Similarly with such terms as deceiving, persuading, even informing, are they permissible only in the Pickwickian senses in which they can attach to genes?

Discussion of whether animal communication is informative or manipulative has assumed that these are exclusive alternatives. Yet, surely, informing could be one way of manipulating; indeed it can be argued that you cannot have one without the other (Staddon, personal communication). But why should we have to suppose that all signals work in the same kind of way? Just as philosophers of language had to shake off a too narrow preoccupation with meaning as naming and describing and turn to consideration of the diversity of uses of language, so students of animal communication may have to give up their preoccupation with information transmission to take full account of the diversity of signal use. Indeed Austin's discussion of "performatives," and its issue in the philosophy of "speech acts" could offer more apt linguistic models than the propositional sentence, which can be true or false, for the interpretation of animal communication.

Study of animal communication, especially in the higher vertebrates, has helped to direct the attention of linguists and other students of human communication to the paralinguistic concommitants which provide speech with context and nuance affecting what it conveys and what it effects (12). This awareness of how human language is interfused with other forms of communication, with convention and cognition, reflects back on views of animal communication as well.

FORMS OF CONVERSE

Perceptual Considerations

Consideration of the channel via which signals are transmitted is basic to understanding of any form of communication. The means of producing signals, the means of their conveyance, and the means of receiving them - expressive abilities, modality properties, sensory capacities - set limits and impose constraints on the possibilities for social intercourse.

Most vertebrates experience the same sensory modalities as we do. Exceptions are certain fishes that are sensitive to electric fields, and some of which generate their own fields, whose deformations they register and respond to as "electrolocation" of objects at a distance. Use of this capacity for production and reception of electric signals for social communication has been experimentally demonstrated. We find it difficult to imagine what experience of such an alien modality might be like, but only a degree more so than imagining what animals experience in sensory modalities that we share with them but from ranges of stimulation to which they are receptive and we are not. Thus our understanding of ultrasonic hearing by a bat is based on "knowledge by description" rather than "knowledge by acquaintance." Even where we may have common range with the animals, their discrimination may be so much coarser or finer than ours as to keep us in the dark about how the world is to them. Even the most expanded human consciousness hardly reaches the threshold of the smell world of a dog. But neither does our having sensory domains and discriminative capacities in common with an animal guarantee that our perception matches the animal's. Especially in studies of communication an observer is likely to find the animals treating things differently or the same, which seem to the observer to be the same or different.

For example, my initial observations of vocal communication in Laughing Gulls took the "long-call" to be a unitary display, the same for all individuals and all occasions in form and function. Variation in response to playback of recorded calls led me to search for and find first, individual differences that serve for individual recognition, then a feature specifying the "address" of a call. This latter, although easy enough to hear once my ear was tuned to it, would, I am sure, never have come to my notice had the birds not shown by their discriminative responses that something had to be there to make the difference (4). Further study has revealed still further complexity, some of which I believe to have semantic and pragmatic significance. This example cautions that we may often misunderstand what animals are doing in social interaction because we fail to draw our distinctions where the animals draw theirs, syntactically and and contextually. Even sophisticated mathematical analysis of such interaction could be idle, at best, unless our description of it corresponds closely enough to that which, so to say, the animals might make of it.

As for what the reactions of animals to social stimuli tell about how social companions are perceived, again the message that one gets most definitely in many cases - which multiply as one goes phylogenetically farther from the human case - is that the animal experience must be different from the human. Tinbergen's sticklebacks displaying to a red mail van; Lack's robins beating on a bunch of red feathers as though it were a rival; my gulls attempting to copulate with a head stuck on a stick; these and cases like them suggest analogy to machines rather than minds - hence the classical ethological concepts of releasers and innate releasing mechanisms. We know that the sensory acuity of the animals is more than enough to register the difference between a van and a fish, a ball and a bird; indeed, at least for the gulls, is even sufficient for individual recognition. The crude stimuli seem ineluctably to overrule the animal's better basis of judgement. Perhaps we get a hint of what this is like when we recoil in horror at the sight of blood, or find ourselves "blind with anger"; but are we ever so uncoupled from our normal perceptual proficiency as to be duped into "miscarriages" of response by the kinds of experiment used by ethologists to probe the perceptual worlds of lower animals? The more these perceptual worlds appear to differ from the human, the less confidently can we conjecture about mental lives centered within them.

On the other hand there is a behavioristic tendency to underestimate an animal's apprehension. White-throated Sparrows not only recognize their neighbors individually by ear, they also know where each belongs (8). When an infant Vervet Monkey screams, bystanders look towards the animal's mother, revealing nice awareness of relationships within the group (Cheney and Seyfarth, personal communication). Such communication behavior implies a surprising degree of "knowing that" (18).

Memory

Individual recognition, dominance hierarchies, territorial habitation imply that animals retain some impressions of one another, past interactions, and local geography. However, memory is seldom investigated directly in the context of communication studies; most of what we know about it comes from the experimental psychology of learning and the physiological psychology of the search for the engram, where I shall leave it.

A point which is seldom considered is that forgetting may be more than a failure to remember. Just as perception has to be selective, lest an animal be trammelled to spasticity by trying to attend to all the stimuli impinging on it, so memory may have to be periodically curtailed lest an animal be so dominated by deposits from its past as to be incapable of outgrowing relationships that have served their turn. Social development in some species involves loss of attachments as well as formation of attachments, and progressive change in the use of and response to signals, which bespeaks flexibility of resource rather than fixity of habit.

Communication in many species is a refined skill, which is acquired and maintained through experience of social interaction. Deprivation of such experience, even for short periods, may have profound and long-lasting deleterious effects on social and cognitive development. On the other hand there are cases of animals raised in isolation from normal social contact which nevertheless showed little if any deficiency in social behavior at maturity. Such cases multiply with phylogenetic distance from the human. Our social and communication behavior is so imbued with effects of experience in society that we can hardly imagine what it might be like to enter social life without them and not be at a loss to know how to cope. Perhaps the best idea we can form of such "untaught knowledge" is to be had from things like blushing, eyebrow-flashing, tongue-showing, which we do most of the time unconsciously, which are not learned in the way our more consciously commanded communication is, yet which betray emotion, behavioral tendency, or situation (12,22). But these are the features of our own communication which carry the least cognitive implication; from which it may follow that creatures limited to such forms of converse have little to say and even less to think about.

Thinking

Surveys of descriptions of communication in vertebrates (20, 21) and theoretical treatments of the evidence (9,22) encourage the view that all animal signals are merely involuntary symptoms of physiological state. Accordingly, Chomsky (5) could argue that in no "known animal communication system...is there any significant similarity to human language." If so, and if thought is tied to language, or something having the discursive and logical capabilities of language, as many assume, then the animals are truly dumb, and their dumbness bespeaks no mind. So, whereof they cannot speak must we be silent?

Langer's (15) thesis about feeling suggest another, if indirect approach. In a study of Japanese Macaques, Green distinguished seven categories of coos, each specifying a different emotional state (9). Eisenberg described comparably complex links between signals and states in Spider Monkeys (9). Green and Marler (9) submit that such cases conform to a scheme devised by Plutchik for representing human emotional states and their expression, and that "Such a hypothetical emotion-based signaling system would provide an economical explanation for much of the signaling behavior of animals, especially if one allows for some degree of mingling of emotional states, and for variations in the intensity of each." However, the situation in some of the more complex cases appears to be that the varieties of expression exceed the varieties of distinguishable emotional state unless one includes the occasioning circumstances in their specification. Consequently, at least some of the signals can convey reference to external objects or the signaler's situation, in addition to expression of emotional state. That a substrate of feeling can go along with ulterior denotation is common experience - feeling is part and parcel of thinking, it "is the mark of mentality" (15).

Some observations more directly suggest that the effective message, as well as the meaning (see Smith (23)) of a signal can be in its external reference. For example, many birds and primates use calls that specify food, in some cases even the quality or desirability of the food (9). Alarm calls provide even more persuasive examples of "symbolic signaling" (9,22). Such evident external denotation does not imply that the calls lack underlying emotional foundation, but since there are unlikely to be differences of emotional state corresponding to each of the situations associated with the calls, the calls arguably specify the situations for the signaler as well as for the receiver.

Contrariwise there are numerous cases, especially among the higher and more social vertebrates, of signals that occur in a range of contexts so various as to defy their being tied to a single emotional state. For example, many of the displays of the Laughing Gull occur in nearly all the motivationally distinguishable situations, implying compatibility with fear, hostility, sexual arousal, urge to incubate, parental solicitude, and other states (3). Smith's (22) solution to the problem of making sense of such cases is that a display is about behavioral tendency rather than emotion or motivation. Thus a call might signify hesitation between moving and staying-put, irrespective of whether the animal is angry or fearful, sexually provoked or parentally harassed, avoiding a predator or maintaining contact with its group. This kind of interpretation, together with the idea that context can add specificity to what a signal means to a recipient, makes more plausible sense of much of the communication data than have some of the more mindless statistical approaches.

However, Smith's view that displays "make information about behavior available" is agnostic with regard to cognitive implications of communication behavior: although leaving open the possibility that the animals know what they are doing, it is as consistent with regarding signals as involuntary symptoms as is the emotional expression thesis. Also, like the emotional expression thesis, it runs into a problem of getting conception and observation to agree in some cases. The contexts of use of a signal can be so diverse and the sequential contingencies so various that if the occurrences can be found to have anything in common at all it may be something so general or vague, something informationally so meager as to seem not worth communicating. An alternative move in such a case is to differentiate according to use or contingency and so regard displays of the same form as informationally distinct.

This distinction between units of form and units of sense certainly applies to the finer divisions that one can make of some communication behavior. The vocalizations of Laughing Gulls comprise a very limited number of types of note, almost none of which signifies singly, but only in sequences differentiated by such characteristics as number and rate of repetitions, selection, and order of notes (4). Analogy with the phoneme/morpheme distinction of speech encourages the conjecture that higher tiers of syntactic hierarchy might obtain as well. In the "meeting ceremony" of courtship the gulls perform agonistic displays in a regular sequence which differs from what happens when the displays occur in territorial disputes and other contexts. The courtship sequences can be read as referring to the fear and hostility, or attack and fleeing tendencies, expressed by the displays in agonistic contexts, but in a manner that denies them. Consistent with this interpretation are indications that one of the movements included in the sequences, "facing-away," in effect constitutes a negation sign (3,4). Experiments with remote-controlled movable and vocal dummies have resulted in some support for the idea that gulls can use the same display to convey different messages by combining or stringing it together with other displays (1). If, as this kind of interpretation implies, gulls can refer to an action in the absence of the motivation to perform it, and do so by using an analogue of grammatical rules, we have to credit them with greater semantic and syntactic sophistication than predicted by either the emotion-expression theory or the behavior-tendency theory of signal production. The cognitive implication is not that a gull is self-consciously aware of referring or rule-following; most of the time we need little if any thought to come up with the right word and use it grammatically (Chomsky even argues that the basic rules of grammar are innate). The same can be said of a monkey who puts on a "playface" to convey that its fighting posture is only in fun, and thus displays "metacommunication." But such performance is hardly conceivable without the kind of "knowing how" that Ryle (18) included in the concept of mind.

The knowing how of competence in a skill suggests itself repeatedly in observations of social interaction among vertebrates. Even as uncerebral a creature as a newt presents an example of choreographic routine combined with countering moves in the display sequences by means of which the male attempts to maneuver the female into snagging the spermatophore he plants for her (11). However, the primates show the most refined facility in their uses of communication. An especially telling example is the duplicity displayed by chimpanzees trying to keep the locality of cached food to themselves, and the cunning of others at beating the bluff (16). Examples like this favor the view that communication is a means of manipulating rather than informing the social companion. Not, to repeat the point, that it must be all one or the other, or could never be both together. Also, by manipulation here I mean the word literally, in the sense of purposeful action, not figuratively in the sense in which it might apply to selfish genes, which make no "conscious prediction" (6). You cannot - logically cannot - tell lies unintentionally; even the idea of self-deception involves the intentional model, one part of the self trying to put it over on the rest. The dissembling chimpanzee appears to be acting on the understanding of what the signs he gives will mean to others, and hence intentionally. A great deal of vertebrate communication behavior is more readily regarded as action undertaken than as movement undergone. If cognition encompasses conation, then a cognitive implication of manipulative signaling is that animals can act with intent.

A development of the conception of communication as manipulation is that it might account for the evolution of aesthetic expression and sensibility (6). The singing of many songbirds is richer in melodic and harmonic content and variation than can be adequately accounted for by its basic biological functions of territorial advertisement and pair-formation, or even by the special explanations proposed for it, such as the thesis that avoidance of monotony counteracts habituation of response, and the Beau Geste hypothesis that birds vary their songs to deceive others into overestimating the number present (14). In some species the songs have formal organization, such as theme and variation, melodic transposition and inversion (24), which appeal to aesthetic rather than functional principles. Whales and some primates are comparably musical in their "singing." The performances bespeak an aesthetic intelligence, and at least suggest that the animals have a sense of beauty and take satisfaction in its indulgence.

CONCLUSION

I have been able to deduce little directly about vertebrate thinking, in a narrow sense, from vertebrate communication; most of what I find the study of vertebrate communication to offer are indirect indications about animal thinking in a broad sense. I argue that a great deal depends upon the investigator's starting stance and preconceptions; that due allowance should be made for the diversity in the types and uses of communication in vertebrates; that while some systems appear to be so different from the human as to be beyond our empathetic apprehension, others appear to involve knowledge, intention, and aesthetic dimension within our ken. By and large the more an animal mind seems continuous with the human mind, the closer the animal to human phylogenetically, and in the texture of its social life. The social insects are exceptions proving the rule: in the sophistication of their symbolic signaling systems, and in the complexity of organization in their societies, they rival the human case as do no other kinds of animal; yet the confined compass of their communication capacities and the impersonality of their social relationships convey an impression of such alien being that we tend to persist in the conceivably mistaken view that they are nothing more than mindless robots. Could we bring ourselves to think otherwise, it would no doubt be on the assumption that "the subtlety and strength of consciousness are always in proportion to the capacity for communication..."

REFERENCES

(1) Amlaner, C.J., and Stout, J.F. 1975. Aggressive communication in Larus glaucescens. Part IV: Interactions of territory residents with a remotely controlled model. Behav. 66: 223-351.

(2) Ayer, A.J. 1968. The Origins of Pragmatism. London: MacMillan.

(3) Beer, C.G. 1975. Multiple functions and gull displays. In Function and Evolution in Behaviour, eds. G. Baerends, C. Beer, and A. Manning. Oxford: Clarendon Press.

(4) Beer, C.G. 1976. Some complexities in the communication behavior of gulls. Ann. NY Acad. Sci. 280: 413-432.

(5) Chomsky, N. 1966. Cartesian Linguistics. New York: Harper.

(6) Dawkins, R., and Krebs, J.R. 1978. Animal signals: information or manipulation? In Behavioural Ecology, eds. J.R. Krebs and N.B. Davies. Oxford: Blackwell.

(7) Dennett, D.C. 1978. Brainstorms. Montgomery, VT: Bradford.

(8) Falls, J.B. 1969. Functions of territorial song in the White-crowned Sparrow. In Bird Vocalizations, ed. R.A. Hinde. Cambridge: Cambridge University Press.

(9) Green, S., and Marler, P. 1979. The analysis of animal communication. In Handbook of Behavioral Neurobiology: Social Behavior and Communication, eds. P. Marler and J.G. Vandenbergh, vol. 3. New York: Plenum.

(10) Griffin, D.R. 1978. Prospects for a cognitive ethology. Behav. Brain Sci. 1: 527-538.

(11) Halliday, T.R. 1975. An observational and experimental study of sexual behaviour in the smooth newt, Triturus vulgaris. Anim. Behav. 23: 291-322.

(12) Hinde, R.A., ed. 1972. Non-verbal Communication. Cambridge: Cambridge University Press.

(13) Hornsby, J. 1980. Actions. London: Routledge and Kegan Paul.

(14) Krebs, J.R. 1977. The significance of song repertoires: The Beau Geste hypothesis. Anim. Behav. 25: 475-478.

(15) Langer, S.K. 1968, 1972. Mind: An Essay on Human Feeling, 2 vols. Baltimore: Johns Hopkins.

(16) Menzel, E.W. 1974. A group of young chimpanzees in a one-acre field. In Behavior of Nonhuman Primates, eds. A.M. Schrier and F. Stollnitz. New York: Academic Press.

(17) Midgely, M. 1978. Beast and Man. Ithaca, NY: Cornell University Press.

(18) Ryle, G. 1949. The Concept of Man. London: Hutchinson.

(19) Schleidt, W.M., and Crawley, J.N. 1980. Patterns in the behaviour of organisms. J. Social Biol. Struct. 3: 1-15.

(20) Sebeok, T.A., ed. 1968. Animal Communication. Bloomington: Indiana University Press.

(21) Sebeok, T.A., ed. 1977. How Animals Communicate. Bloomington: Indiana University Press.

(22) Seyfarth, R.M.; Cheney, D.L.; and Marler, P. 1980. Monkey responses to three different alarm calls: evidence of predator classification and semantic communication. Science 210: 801-803.

(23) Smith, W.J. 1977. The Behavior of Communicating. Cambridge, MA: Harvard University Press.

(24) Thorpe, W.H., and Hall-Craggs, J. 1976. Sound production and perception in birds as related to the general principles of pattern recognition. In Growing Points in Ethology, eds. P.P.G. Bateson and R.A. Hinde. Cambridge: Cambridge University Press.

Animal Mind - Human Mind, ed. D.R. Griffin, pp. 269-298.
Dahlem Konferenzen 1982. Berlin, Heidelberg, New York: Springer-Verlag.

The Insect Mind: Physics or Metaphysics?

J.L. Gould and C. G. Gould
Dept. of Biology, Princeton University
Princeton, NJ 08544, USA

Abstract. When we attempt to infer from an animal's overt behavior whether its brain might be thinking or merely computing, a variety of intuitively suggestive lines of evidence become unreliable. Many behavioral traits such as adaptive behavior, behavioral variability, complexity, flexibility (including learning), and even the phenomenon of culture can be - and in insects at least have been proven to be - the results of genetic programming. The demonstrated effects of programming can be so intricate and subtle that even what seems to be insight or creativity must be suspect. A versatile and more reliable guide to the inner workings of minds is communication. Experimental manipulations of signals can show, through their effects on the behavior of the receiver, how the incoming information is being processed. This line of inquiry has laid open much of the insect mind, particularly that of the honey bee, but no compelling evidence for awareness has emerged. Instead, insects stand more than ever as testaments to the power of blind behavioral programming, and as such remind us to be wary of attributing to vertebrates anything more than larger, more interesting on-board computers.

THE PROBLEM

The sources of the knowledge and the motivations which guide the behavior of animals continues to provoke interest and controversy. Spalding, for example, marvelled at the seemingly prescient behavior of the female wasp who would labor tirelessly to "gather food...she never tasted" to feed "larvae she would never see" (49). The remarkable intricacies of the wasp's behavior must result either from instinct or from learning: the information necessary for digging her tunnels, for hunting and paralyzing her one species of prey, for navigating back home, and for laying an egg and sealing the chamber after having collected the correct number of victims is either innate - coded for by the genes as "instinct" - or learned through experience.

Even the most determined advocates of the role of nurture over nature now generally concede that the clockwork behavior of wasps and other insects is largely prewired. But when it comes to higher animals - humans in particular - many of us are more reluctant to invoke the notion of instinct to explain at least certain examples of seemingly intelligent, adaptable, and highly complex behavior. But the mind is a private organ whose inner workings must normally be inferred from observation of the overt behavior it directs. When we watch an animal, how are we to know whether it is consciously thinking or merely computing; whether it is a sentient being, or simply an unconscious, well-programmed robot; whether the knowledge which guides behavior represents the intelligence of the individual, or of evolution? Since the question of what is going on in the brains of animals, whether encumbered by backbones or not, must usually be approached so indirectly, what are the crucial, telltale signs which might indicate a glimmer, however faint, of self-awareness, self-direction, and everything else that "animal consciousness" implies?

ADAPTIVE BEHAVIOR

One possible argument in favor of awareness is that animals regularly face difficult situations and resolve them in an intelligent way. Perhaps this implies that they have some "intellectual" grasp of the problem. A goose, for example, spotting an egg which has unaccountably managed to escape from the confines of her nest will stare at it, rise, extend her neck over the egg, and roll it gently back into the nest with her bill. Does this thoroughly sensible solution to what must be a fairly rare contingency represent the goose's personal evaluation of the problem? Alas, as Lorenz and Tinbergen showed, the egg-rolling response is a mindless bit of programming based on an innate recognition circuit which triggers a prewired, centrally coordinated motor response. The recognition is accomplished on the basis of sign stimuli - crude but normally diagnostic features of the appropriate stimulus. Despite having incubated her eggs for days, regularly examining them for accidental breaks, and rolling them to keep the embryos from sticking to their shells, the goose has no real idea of eggs and will gladly retrieve a motley collection of ping-pong balls, batteries, and beer bottles. The motor program by which the egg is recovered is equally mindless, so that removing the egg after the goose begins to reach for it does not prevent her from gingerly rolling the nonexistent egg back into the nest (34).

The ethological principles of sign stimuli (key features which may have no particular salience for other species) and motor programs, as well as the discovery that many species are tuned to sensory worlds outside of our own experience, have served to clear up much of the superficial mystery surrounding behavior. The subsequent realization that sign stimuli for the most part correspond to the optimal triggers for specific feature-detector cells in the nervous system (22) combined with the painstaking mapping of motor-program circuits on a cell-by-cell basis (50) underscore the mechanical basis of most of the behavior of most animals. This is not to say that the existence of mindless programming in one aspect of an animal's behavior rules out the possible role of consciousness in another, but it is clear that evolution's ability to prewire simple but adaptive behavioral units renders the commonplace examples of apparent problem analysis and adaptive behavioral responses nearly useless measures of awareness.

VARIABILITY

Though in most contexts many animals may be nothing more than preprogrammed robots, perhaps the very regularity and invariability of their internal clockwork might serve as a guide to consciousness. After all, though intellect will often serve up two different answers to the same question asked at different times, good machinery is much more reliable. For example, when Fabre interfered with the prey-capture ritual of a cricket-hunting wasp by moving the paralyzed victim (which the wasp invariably leaves lying on its back, its antennae just touching the tunnel entrance, while she inspects her burrow), the re-emerging wasp insisted each time on repositioning the cricket, and inspecting her tunnel again. Fabre continued this trivial alteration 40 times and the wasp, locked in a behavioral "do-loop," never thought to skip an obviously pointless step in her program (11). Clearly the wasp is a machine in this context (and in many other examples described by Fabre), entirely inflexible in her behavior. But what would variability have told us? Fabre found another population of this cricket-hunting species that eventually would skip the repositioning maneuver and take the misplaced cricket down inside. Have these particular wasps evolved consciousness, or were the gears of their brains simply less well made?

The remarkable nature of the wasp's performance serves to remind us that most other animals have contingency plans to extricate them from such behavioral cul-de-sacs. By far the most common escape mechanism is habituation, a central sensory or behavioral boredom by which the level of responsiveness to a stimulus decreases with its continued presentation. Hence a goose may recover the first egg of the day with admirable parental promptness, but by the tenth the goose grows quite reluctant. But habituation is merely a programming ploy, a piece of behavioral machinery whose neural and biochemical bases are even now beginning to yield up their secrets (7,28), and the slow and generally logical development of this behavioral numbness may be swept mindlessly away by any of a variety of irrelevant stimuli through the familiar process of sensitization.

Other sorts of seemingly intelligent behavioral variability, though, cannot be accounted for either by "noise" in the computer or by habituation. Although geese and gulls, for example, quite sensibly roll eggs during incubation, the egg-rolling response and a separate program for removing damaged or broken eggs switch on three weeks before egglaying and turn off well after hatching (1). The survival value of the shell-removal behavior is clear: the white interior of a broken egg destroys the nest's camouflage (52). This intelligent-looking behavior is utterly mindless and depends on crude sign stimuli: parents will remove inconspicuous bottle caps, eye-dropper bulbs, and folded brown paper; while eye-catching white batteries, skulls, and stones will be left in the nest. Obviously, these birds have no idea of eggs, camouflage, or timing.

It seems clear that this orchestration of subroutines - a phenomenon often referred to as "drive" - is just another bit of clockwork. The usual distinguishing characteristics of drive are that it makes behavioral units available before their specific triggering stimulus appears, and that it makes liberal use of internal timers. The logic of this system is that many behaviors require time-consuming physiological preparations. Ring doves, for example, produce the crop milk which they feed their young 16 days after they see the eggs in their nest: it is the <u>sight</u> of the eggs, rather than the egglaying process itself, which is the trigger (30). The preparations for the production of the crop milk are accomplished by a hormone, prolactin, which acts as an internal pheromone. Similarly, the preparations and behavioral changes which occur in song-birds prior to staking out territories (such as development of the song, the appearance of sexual plumage, and aggressiveness) are triggered by testosterone, its production triggered in turn by the increasing day length which presages the coming of spring (12).

Although sensible long-term changes in behavior have proven most often to be the consequences of programming rather than thought, what about short-term variability? Honey bees, for example, show spontaneous preferences for certain colors and shapes of artificial flowers (24). This display of aesthetic preference is not absolute, but rather probabilistic: in a two-color choice test between hues reliably distinguished in learning experiments, color A might be chosen only, say, 60% of the time rather than 100%. Similarly, in a conflict situation an animal will sometimes fight and sometimes flee. Even in experiments in which care has been taken to factor out the role of immediate past experience, this sort of quantitatively predictable variability persists.

Can the perplexing unreliability of animal behavior be taken as evidence for something more (or, perhaps, less) than machinery making decisions? Alas, probably not. The probabilistic strategy of decision making is evident even in bacteria (17) and has been given a firm theoretical foundation by Smith (48) and others based on game theory. In short, it is often most adaptive to be variable or unpredictable, so long as evolution or personal experience takes care to set the odds appropriately. Hence, though flowers may more often be color A than B, it makes more sense to try B-colored objects from time to time rather than to concentrate exclusively on color A. There seems little reason, therefore, to invoke consciousness to account for the behavioral fickleness of honey bees or, indeed, of animals in general.

COMPLEXITY

Although a great deal may be programmed into animals, there must surely be a limit to the complexity possible. There must be a point beyond which no set of computer-like elements can suffice to account for an animal's apparent grasp of its situation, particularly in the face of variable or unpredictable environmental contingencies. The difficulty in drawing this intellectual line, however, is great. Some of the most impressively complex examples of behavior we have are known to be wholly innate. The nest building of weaver birds is an apt example. Their intricate nests, involving as they do thousands of knots whose location and type depend on utterly unpredictable contingencies of leaves, boughs, material, and prior knots, seem well beyond the scope of any programming we can imagine. And yet there they are, fabricated perfectly well without any need for previous experience or observation on the bird's part, testimonies to the poverty of our imaginations and current understanding. How are the many different subroutines orchestrated to fashion the support, the nest bowl, the cover, and the long entrance tube? Does the bird come with a mental picture of the finished product to match as best it can, or with a series of steps to be accomplished, or with a knot-by-knot score to follow?

Given the undoubtedly limited intellectual ability of the performer, surely the construction of orb webs is even more impressive. In total darkness, without prior experience, and with the location of potential anchor points for the support structure unpredictable, a mere spider sets about constructing a precise and complex network of several different kinds of threads. Even damage during construction is automatically repaired. Doubtless all this is accomplished through one master program and several subroutines and requires no conscious grasp of the problem.

The use of subroutines to deal with the unpredictable is especially obvious in navigation. Honey bees, for example, regularly use the sun as their compass, compensating for its changing azimuth as it moves from east to west. Their compensation is perfectly mindless, depending upon a memory of azimuth relative to the bees' goal on the previous trip (or day) and an extrapolation of the sun's current rate of azimuth movement (21). The sun is recognized by an equally mechanical sign stimulus - its low ratio of UV to visible light - so that a dim 10°, triangular, highly polarized green object against a dark background is just as acceptable to bees as the actual intensely bright $1/2^{o}$, circular, unpolarized white sun in the normal sky (4).

When the sun is removed (obscured perhaps by a cloud, a landmark, or the horizon), the whole sun-centered system of bees is discarded in favor of a backup system - a separate navigational subroutine - based on the patterns of polarized light generated in the sky by the scattering of sunlight. This analysis itself is composed of a primary and backup system and uses sign stimuli and very simple processing. When the polarization too is removed (as on overcast days, for example), bees fall back on yet a third system. The apparent complexity of the formidable navigational task of even insects is, in fact, based on the interplay of a group of subroutines which are themselves quite simple, depending on the same sorts of schematic stimulus-recognition systems and relatively uncomplicated processing seen in less elaborate behavior. In short, a staggering degree of behavioral complexity can be generated by a set of individually simple subroutines, and the result is that mere complexity of behavior is no trustworthy guide to awareness.

LEARNING

Another common strategy for dealing with the unpredictable contingencies of an animal's world is learning, and it is here that our intuition tells us that we must be dealing with something very near consciousness. After all, does not learning suggest some degree of understanding, a conscious comprehension of the problem to be solved? Alas, even headless flies can learn to hold their legs up or down to avoid a shock, and do so faster than those still encumbered with brains (6). Moreover, there is much in the organization of invertebrate learning which suggests the gears and wheels of an automatic pilot rather than any aware intelligence. When a honey bee discovers a flower, for example, she sets in motion a learning sequence which seems utterly mechanical in nature. A forager learns many things about a food source which aid her in the future, including its color, shape, odor, location, nearby landmarks, and time of nectar production. As Opfinger discovered (40) and Menzel and his associates have detailed (39), the color is learned only in the final three seconds as the bee lands. The colors clearly visible to the bee before the landing, while standing on the source to feed, and while circling the source before flying on, are simply not remembered. A naive bee carried to the feeder from the hive will circle the source repeatedly after taking on a load of sugar water, and yet when she returns a few minutes later she will be unable to choose the correct color. And yet, so mechanical is this learning routine that if we interrupt such a bee while she is first feeding so she must take off and land again on her own accord, that landing permits her to choose the correct feeder color on her next visit. Similarly, bees learn landmarks after taking off: a recruit which arrived, inspected the feeder, but was transported back to the hive while feeding, returns without the slightest memory of the landmarks she must certainly have seen on her arrival.

Other aspects of flower learning seem equally curious. Although a bee learns a flower's odor almost perfectly in one visit, several trips are required to learn its color with precision, and even then a bee never chooses the correct color 100% of the time. Time of day is learned more slowly still. It is as though perfection, though obviously possible, is not desirable, since the speed and reliability of a bee's flower memory at least roughly corresponds to the degree of variability it is likely to encounter among flowers of the same species in nature (including variation from day to day of an individual blossom).

Once a bee has learned about a source, the information seems to be stored as a set in the manner of an appointment book. Hence, changing any component of the set - the odor, say, which is relearned to virtual perfection after one visit - forces the bee to relearn painstakingly all the other pieces of information at their characteristic rates even though they have not changed (5). It seems clear that bees are well-programmed learning machines, attending only to the cues deemed salient by evolution (and then only in well-defined contexts and often during precise critical periods) - and then filing the information thus obtained in preexisting mental arrays. Nothing in this behavior, wonderful as it is, suggests any true flexibility or awareness.

How different are vertebrates? Although memory structure seems to be similar (both groups have long- and short-term memory phases (39,43) as well as an assortment of mental functions formally named by learning theory, such as "cognitive maps" (27)), perhaps insects lead such simple and routine lives that their learning can be organized by rote, while higher vertebrates must actually appreciate the problems posed by their complicated worlds, and so exercise some self-directed choice in learning. Although this view may be intuitively appealing, examples abound of programmed learning in vertebrates. The classic case, of course, is parental imprinting, the process by which precocial young learn to recognize their parents (22). Their learning routine begins with the "following response," a behavior by which a duckling, say, follows virtually any moving object producing the species-specific exodus call. The following itself triggers the learning which leads to the ability to recognize the parents as individuals. (Passively transported chicks, like bees ferried to a feeder, fail to learn.) This learning typically takes place during an early and relatively brief critical period and is so mindless that birds may easily be imprinted on inanimate objects such as balls and shoe boxes.

Another lesson of vertebrate learning is that animals often cannot learn what they are not specifically programmed to learn. Birds such as penguins which nest close together or species which are subject to nest parasites memorize the exact appearance of their eggs, while others such as herring gulls pay no attention to even dramatic changes in egg number, size, or color. Once the chicks of ground-nesting gulls hatch, the parents imprint on their young while the young are imprinting on the parents. So exact is this memorization of what seem to be indistinguishable chicks that adult gulls will regularly reject and even eat any of their neighbors' young which might happen to wander in. The cliff-nesting kittiwake gull, however, does not face the contingency of unrelated chicks happening into the home grounds, and so, not being programmed to regard its chicks carefully, will accept even large black chicks in its nest of small white birds (9,51).

Probably the most complex case of programmed learning dissected to date is the song learning of sparrows (36,38). Although these birds do learn their songs, they will learn only the song of their own species, and then only during a specific sensitive period. They recognize the song by one or two schematic criteria common to the variety of songs produced by the species, and they store it for several months before the males use it as a model for song learning and the females for song recognition. Deafening experiments demonstrate that once song has crystallized, the resulting motor pattern becomes as hardwired as any innate motor program. Again this wholly innate orchestration of learning provides for an apparent flexibility and variability, but there seems no reason to suppose that individual birds appreciate what they are about.

The programming of learning, then, serves to direct an animal's attention to a particular stimulus at a particular time in a confusing world full of potentially learnable stimuli. Learning of this sort seems to be a subroutine, called into play when the genes are unable to encode recognition of a stimulus with sufficient precision either because it is too complex to be specified by sign stimuli (though sign stimuli are used liberally to direct the learning at an appropriate target) or because the stimulus is not sufficiently predictable. Hence, though a UV-dark center/UV-bright surround pattern indicates a flower to a bee, or a moving spot and vertical bar encode the essence of a parent to a newborn gull, only directed learning can enable either animal to memorize enough about the appropriate stimulus to recognize it in a choice situation later on. In short, the innate releaser directs the learning of a new releaser or motor program which takes on all the characteristics of an automatic, hard-wired program.

The role of programmed learning in the lives of animals (including us) is almost certainly underestimated because of our too-ready willingness to imagine that animals are aware enough to decide for themselves what to learn. Curio's recent experiments on cultural transmission of information in blackbirds is a case in point (9). Birds in one cage were incited to attack an innocuous stuffed bird or even a milk bottle by the mobbing calls and the behavior of birds in another cage which were permitted to see an owl. The birds indoctrinated against milk bottles could then pass their mindless aversion on to other birds. The mobbing call, of course, is the sign stimulus which directs and triggers this learning and which accounts for the perpetuation of this element of culture.

INSIGHT AND CREATIVITY

Although the widespread instances of programmed learning are doubtless ploys by the genes to tell their dim-witted couriers when and what to learn and then what to do with the knowledge, cases of apparently self-directed learning may admit of another explanation. When the first blue tit discovered how to open a milk bottle to get at the cream (25), or the Japanese macaque named Imo hit upon washing the sand off of sweet potatoes (29), a new element was introduced into their respective cultures. Now although the evidence for programmed copying as the basis for the cultural transmission of food rituals in both birds and primates is suggestive (37), the original discovery could have involved more than simple trial and error. The rarity of innovation, combined with the relative rapidity of copying, argues for the dumb-luck explanation; while the subsequent discovery by the very same macaque of a method to separate wheat from sand suggests the possibility of true insight, a sort of cognitive trial and error which would seem to require an appreciation of the problem.

Although it is tempting to infer the existence of some sort of awareness in primates on the basis of this sort of evidence (not to mention the mirror experiments of Gallup (15), the playback of alarm calls to vervets (47), and the successes of the Gardners (16), Rumbaughs (45), and Premack (42) in teaching language of a sort to chimpanzees), there are similar cases in animals for whom the possibility of consciousness seems plainly absurd which suggest as powerfully the grasp of a problem. One example revolves around the avoidance of alfalfa by honey bees. These flowers possess springloaded anthers which give honey bees a rough blow when they enter. Although bumble bees, which evolved pollinating alfalfa, do not seem to mind, honey bees, once so treated, avoid alfalfa religiously (35). Placed in the middle of a field of alfalfa, foraging bees will fly tremendous distances to find alternate food sources. Modern agricultural practices and the finite (though surprisingly long) flight range of honey bees, however, often brings bees to a grim choice between foraging on alfalfa or starving.

In the face of certain starvation, honey bees are said finally to begin foraging on alfalfa, but they learn to avoid being clubbed (41,44). Some bees come to recognize tripped from untripped flowers and frequent only the former, while others learn to chew a hole in the back of the flower so as to rob untripped blossoms without ever venturing inside. Who has analyzed and solved this problem: evolution, or the bees themselves? Perhaps evolution has designed bees with a contingency plan for American agriculture, or it may be that both cases are standard, pre-wired ploys - chewing through being reserved normally for robbing flowers too small to enter, while differentiating tripped from untripped blossoms could be simply a far more precise use of species differentiation.

Another, slightly eerie case is not so easy to dismiss. During training to an artificial food source, there comes a point at which at least some of the bees begin to "catch on" that the experimenter is systematically moving the food farther and farther away, and von Frisch recalls instances in which the trained foragers began to <u>anticipate</u> subsequent moves and to wait for the feeder at the presumptive new location (14). If anything, this seems a more impressive intellectual feat than potato washing. The argument that this behavior is an artifact of the bees' ability to extrapolate azimuth movement (21) is weak and unconvincing, and it is not easy to imagine anything in the behavior of flowers in nature for which evolution could conceivably have needed to program bees to anticipate regular changes in distance.

COMMUNICATION

If we are to follow our intuition and deny even a dim awareness to honey bees despite their seeming ability to grasp certain unusual problems, where does this leave us in regard to primates or, for that matter, our own species? The problem here is, of course, the one we began with: how are we to determine with any certainty from the overt behavior of an animal exactly what is going on in its mind? Griffin suggests that the best mental window is communication - not observing it passively, but actually entering into it (23). Of course this is simply an adumbration of the highly successful technique pioneered by Tinbergen, by which ethologists have been able through model experiments to isolate from all the potentially useful information broadcast by a sender those few, essential features which convey meaning.

The role of communication in most nonsocial species (and virtually all species are nonsocial except during courtship) is rather simple, and not at all suitable for our purposes, since it usually imparts merely a species-specific message indicating the sex and reproductive readiness of the sender. In relatively social species, however, communication takes on the more complicated tasks of indicating mood and status, and facilitating social coordination and control. A little discreet intervention in such systems can tell us a great deal about any mental processing which may be going on in the actors. For instance, social insects regularly employ a variety of chemical signals which we anthropomorphically label attraction or recruitment pheromones, alarm substances, panic odors, and so on. Are insects aware of the meaning of these messages in any real sense, or are these simply sign stimuli which trigger automatic responses?

Except for cases in which our current ignorance permits us to imagine otherwise, the latter interpretation seems most appropriate. For example, a honey bee provides its last social message almost the moment it dies in the form of a death pheromone: oleic acid. Other bees then remove the corpse from the hive. Their response to this bit of communication is entirely innate and utterly mindless: a drop of oleic acid painted on a live bee, even the queen, will result in the unfortunate and obviously healthy insect's being carried kicking and screaming out of the hive. Similarly the alarm pheromone, combined with a bit of rapid side-to-side movement, will induce foragers near the hive to attack a food source they have been collecting from for hours. Then too, ants regularly accept symbionts and parasites in their midsts, creatures with grossly different anatomies, simply because they transmit the appropriate chemical codes (26). Social caterpillars will mindlessly follow a small circular trail round and round until they drop. In sum, there is nothing in the behavior of insects to suggest that they have any real comprehension of the meaning or the circumstances surrounding the chemical messages they use.

Of all the examples of social communication in insects, the most impressive and most likely to reveal any understanding on the part of the participants is the dance language of honey bees. A forager returning from a good food source will often perform a dance which von Frisch showed encodes the distance and direction of the food (13). The dance is normally performed on vertical surfaces of honey comb in the darkness of the hive. The distance to the food is specified by the number of waggles in the straight-run segment of the dance, while the direction is indicated by the angle of the run relative to vertical. Recruits use the information gained from the dance to find the food (14,19,20). This extraordinary communication system is called a "language" because it refers to objects distant both in space and time and depends upon arbitrary linguistic conventions. For example, the sun is taken as the primary reference point (presumably the bees could have used north or the direction the hive is facing or any other direction) and is defined as "up" on the comb (whereas "down" or, for that matter, 13° left of vertical could have been selected). The distance system requires defining the value in meters for each waggle, and the arbitrary nature of this convention is tellingly illustrated by the existence of wildly different conventions - dialects - among different races of bees (3). Two more dance conventions have been discovered recently (4).

The dances of forager bees serve as windows into their minds, telling us how they have construed the information collected during their journey. In one of a remarkable series of experiments, for example, von Frisch and his colleagues showed that bees integrate over an indirect path around obstacles such as mountains or buildings so as to indicate the true direction of the food (14). This elegant piece of computation is well adapted to get a forager home after a circuitous search, but, as the dancer ought to know, is not at all helpful to recruits when obstacles are involved. The tendency of many foragers on trip after trip to stick with an indirect route out and a direct path back also reflects adversely on bee intellect.

In another set of experiments, von Frisch showed that distance is judged by effort, so that foragers carrying weights or drag-producing flaps overestimate distance (as, indeed, do bees constrained to walk to food). Moreover, the distance estimation is based on the outward flight: bees overestimate the length of the journey when the trip out is into the wind or uphill, whereas it is underestimated when the goal is downwind or downhill (14). This system results in a forager's grossly overestimating the distance on indirect routes even though the flight back to the hive may have been direct, and so wrongly informs the recruit who is being told of the true direction. The only exception to this effort-as-distance picture is an experiment in which von Frisch trained bees up a 75m tower - about 3 waggles away for horizontal flight - and although the effort required for going up is obviously much greater than for flying horizontally, the bees performed the zero-waggle round dance normally reserved for food within about 20m. Another curiosity is that foragers do not really signal the azimuth of the source directly. When the food is moved with respect to the sun, dancers having just flown to and from the source in the new direction nevertheless report the former azimuth. Only gradually, over a period of 30-45 minutes (about 5-10 round trips) do they begin getting the direction right (33). This seems to imply a mindless running-average strategy for dealing with the sun's movement (21).

The most useful way to probe the dancer's mind is to force her to dance on a <u>horizontal</u> surface. Deprived of the gravity which is her normal orientation cue, she uses instead any visual cues that might be overhead and aims her dance directly at the food. This is not at all an unnatural situation for honey bees. The two tropical species, dorsata and florea, live on exposed comb hung from tree limbs and dance only on horizontal surfaces. Our species, mellifera, shares a common ancestry and still often dances on horizontal surfaces outside the hive under normal conditions and on the cluster itself during swarming. Hence we can tamper with the system by providing artificial celestial cues and see how the dancers interpret them. It is this technique which revealed the bees' use of a sign stimulus for the sun, as well as their arbitrary rules for deciding which of two possible interpretations to put on a pattern of polarized light (4). In the latter case, other cues exist which would help make this determination accurately, but the bees ignore them in favor of a mindless intellectual short circuit which, though technically wrong about half of the time, produces consistent and self-cancelling errors.

The dance also tells us something about honey bee economics. The vigor and duration of the dancing indicates something about the quality of the source, though exactly whose judgement is involved is not clear. Although a forager who is able to distribute her load quickly to waiting bees is far more likely to dance than one who has difficulty (31), the forager factors in her own "opinion" of her goods as well. Recruits also make judgements, comparing different dances before flying out. In both cases distance is a major variable in the calculations: to be acceptable a source far away must offer more concentrated nectar than one near by (2). Since net calories obtained per unit time are the ultimate measure of quality, the cost in terms of fuel, wear and tear, and time in getting to a source is a crucial element in computing efficiency. The judgements themselves are relative, since on a day when lots of food is available even a 2M sugar solution will go unadvertised, while on a bad day food one-tenth as sweet may elicit frantic dancing and recruitment. Oddly enough, however, difficulty in obtaining food actually increases the dance vigor. Using feeders with tiny capillaries, solution artificially thickened, and even sugar water spread on filter paper - operations each of which dramatically reduce the efficiency of foraging by increasing the time required to fill up - have the perfectly illogical effect of enhancing a source's subjective value (14).

The dance also tells us something about the swarming economics of bees. Scout bees search for potential nest sites and inspect them. First Lindauer (32) and later Seeley (46) looked for preferences through choice experiments. Scouts take into account the size, height, dryness, draftiness, and exposure of the cavities they locate, and advertise their best finds. They also visit each other's discoveries and each individual forms a preference. Scouts reexamine the possibilities from time to time so that altering a favored cavity by, say, pouring water on its floor, can affect its popularity radically. There is also a distinct distance and size preference which we have found to differ between races in an ecologically and evolutionarily sensible way. Although the choice and judgement behavior of scouts seems remarkably intelligent, no concerted effort has been made to fool the scouts in order to determine whether or not this is just another bit of clever programming.

DIRECT EXPERIMENTS

In an effort to challenge the bees, who seem to us to be the epitome of insect intellect, with conditions which might reveal any real awareness of their situation, we have planned or tried in at least a preliminary way five approaches. The first and most obvious has been to construct an artificial dancer so that the sender's message can be manipulated. Esch took on this task and, though he learned much about bee communication in the process, he was unable to recruit bees (10). Working from Esch's discoveries, we have made some further progress, but have run into another set of technical difficulties with regard to reproducing the dance sound (18).

The second approach has been to see what happens when a forager offers a message which is obviously untrue. We made use of the "misdirection" technique by which foragers can be made to reference their dances to gravity while recruits interpret them with regard to an artificial sun (19). The usual result of this use of two different direction dialects is that most recruits go to the imaginary source. In an attempt to provide a transparent lie for the dance attenders to evaluate, we trained foragers 165m along the shore of a lake and misdirected the dances to indicate a spot in the center of the lake. The dances were well attended and many potential recruits left the hive, but the recruits stopped at the edge of the lake and searched the shore. Many bees attended these dances repeatedly, each time engaging in a fruitless search of 4-8 minutes, apparently none the wiser for their experience. (Indeed, there seems no reason to suppose bees remember anything from dances which have recruited them previously, even if only

5 minutes earlier (19).) Eventually, a few recruits found the station on the shore (though a substantial proportion of these had not observed a dance), while no bees arrived at a feeding station on a boat at the indicated spot. Recruitment when dances were not misdirected was extraordinarily heavy.

Our first guess was that recruits might be rejecting the lake on the basis of "locale odors" carried (along with floral scents) on the waxy hairs of the foragers. Recruits regularly use food-source odors in their searches and seem to use locale odors at least as a backup to the dance. However, when we trained the foragers to the boat and did not misdirect the dances, recruits found neither the unforaged station on the shore nor, on five of the six days, the boat. (A strong on-shore wind may account for our one success.) If the lack of recruitment cannot be blamed on locale odors, what about simple hydrophobia? Unfortunately, when the boat was taken to the opposite shore 200m across the lake recruitment began again. Did these recruits get there by flying over the lake they otherwise avoided, or by going the 1000m around the edge of this oblong body of water? Almost certainly they flew over. Curiously enough, though the foragers stayed with the boat on its slow trip from the center to the far shore, they abandoned it 5m offshore on the way back and waited for it at the old spot in the center of the lake.

The lake experiment results, though somewhat confusing, offer little support for a theory of bee consciousness. They do, however, suggest that bees might have a mental map of their surroundings. Indeed, the ability of experienced bees to fly directly home after having been caught at the hive entrance and carried several hundred meters away argues for such a map; and the ability of foragers to fly straight back after a circuitous search for food would be more easily explained in this way than on the basis of ongoing trigonometric calculations. The lake experiments, then, may have something to say about the nature of mental representation - cognitive maps - in bees.

Beyond the lake experiments, we hope to give the bees three other situations to analyze. In the first, we will train foragers around a large circular stadium 20m high using only those bees which refuse to fly up and over to return home. As we close the circle, we wonder at what point if any the foragers will realize that a shorter return path exists. If bees do have a mental map, this experiment should shed some light on it. In the second, we will try to document more precisely the behavior of foragers anticipating the move of a feeder.

In the third, we plan to look more closely at the curious "noise" in the dance. Foragers regularly give distance and direction values scattered about the true mean (with an SD of about 30m), even though in a featureless environment their accuracy in finding the location of a station removed during their absence is about 1m. Nor does the inaccuracy, at least in the indication of direction, represent any difficulty in measuring angles in the hive since the angular error virtually disappears in dances for greater distances and is about the same for both vertical and horizontal dances. Two possible explanations are known: the scatter might represent the noise in the time-averaged input of navigation information on the trip (unlikely, we think) or, since the absolute scatter is relatively constant with distance, this equivocation might represent an evolutionary strategy to spread out the recruits to cover what is normally a resource which exists as a patch. In this case we would expect our aware forager to generate different dance scatters for different sources, minimizing it especially in the case of potential nest cavities.

CONSCIOUSNESS

In considering the issue of mental experiences in animals, we have begun to wonder if the implicit assumption that humans are almost wholly conscious and aware (and hence fully competent to evaluate our cognitively less sophisticated animal brethren) is correct. Could it be that the degree to which conscious thinking is involved in the everyday lives of most people is greatly overestimated? We know already that much of our learned behavior becomes hardwired: despite the painfully difficult process of learning the task originally, who has to concentrate consciously as an adult on how to walk or swim, tie a shoe, write words, or even drive a car along a familiar route? Certain linguistic behavior, too, falls into such patterns. Michael Gazzaniga, for instance, tells the story of a former physician who suffered from a left (linguistic) hemisphere lesion so serious that he could not form even simple three-word sentences. And yet, when a certain highly touted but ineffective patent medicine was mentioned, he would launch into a well-worn and perfectly grammatical 5-minute tirade on its evils. This set piece had been stored on the undamaged right side (along with the usual collection of songs, poetry, and epigrams) as a motor tape requiring no conscious linguistic manipulation to deliver.

Such instances are seen frequently in men and animals. Learning can be (and most often is) used as an intellectual shortcut to make the once-plastic rote, to weed out as quickly as possible any need for conscious thought which might otherwise distract that most delicate of human faculties from genuinely new and difficult problems. In such instances only an arbitrary subset of the cues which accompany the appropriate stimulus and context is committed to memory and serves in the future as a sort of sign stimulus for triggering and directing the behavior.

Evidence from "split-brain" patients, however, argues that consciousness is not entirely devoted to creative problem solving. The split-brain operation severs the cerebral commissure, the enormous nerve tract which connects the left and right halves of the brain, and so leaves them unable to communicate with each other. Since the left hemisphere receives its input from the right hand and right visual field, and <u>vice versa</u>, it is possible to show a picture or word to one hemisphere and ask the hand under the control of the other side to select the corresponding object out of a box on the basis of feel alone. Split-brain subjects will gamely search and select an object, but their choice is almost always wrong. When faced with this discrepancy between intent and action and asked why they have made the wrong choice, subjects respond with a wide range of sincere but transparently false rationalizations. A series of such obvious mistakes leaves the subject agitated, but he can be calmed by the experimenter by being reminded of his operation and the artificial nature of the tests. Thus reassured, the subject continues with the experiment but persistently produces the same fanciful explanations for mistakes without ever referring to the surgery.

In short, subjects resolve their intellectual dilemma through a quick, thoroughly unconscious fabrication. If different pictures are shown to a split-brain patient's two hemispheres simultaneously and the subject asked to choose appropriate matches from a set of pictures, each hemisphere will direct its hand independently of the other to its own logical choice - yet only the verbal side will be able to explain its actions. If, as in one of Gazzaniga's experiments, the verbal side sees a bird's claw while the nonverbal hemisphere sees a snowy landscape, one hand will select a picture of a chicken and the other a photograph of a snow shovel. Asked why he chose as he did, the subject's linguistic hemisphere will respond unhesitatingly that the claw of course belongs to the chicken, and the shovel is for cleaning up after it. Confronted with a nonsensical situation, his brain strives valiantly to impose some order on it, to make comprehensible the inherently senseless.

This automatic and widespread drive to rationalize the inexplicable is well-known in the psychological literature as "confabulation" and "cognitive dissonance" and is obvious in our species from the first moment an infant can string together enough words to manage it. Indeed, it seems likely that our need to justify our own behavior and that of the world about us, combined with our facility for divining patterns, real or imaginary, provides the creative basis of art, science, and religion. But to what extent can we call this sort of imaginative self deception "consciousness"? Indeed, what evidence is there to suggest that those sublime intellectual events known as "inspiration" involve any conscious thought? Most often our best ideas are served up to us out of our unconscious while we are thinking or doing something perfectly irrelevant. Inspiration probably depends on some sort of repetitive and time-consuming pattern-matching program which runs imperceptibly below the level of consciousness searching for plausible matches.

It strikes us that a skeptical and dispassionate extraterrestrial ethologist studying our unendearing species might reasonably conclude that Homo sapiens are, for the most part, automatons with overactive and highly verbal public relations departments to apologize for and cover up our foibles. We wonder if our own anthropocentric biases might not combine with the devious and self-serving nature of much conscious thought to mislead us not only about the degree of conscious control we exercise over our own behavior, but the essential role self awareness plays in humans. It seems to us that consciousness serves two main functions in our species: it permits us to solve unexpected problems by formulating novel hypotheses, and it allows us to lie plausibly. Viewed in this light, the true litmus of consciousness may be the ability a) to recognize a logical conflict between a token (the sign stimulus, whether prewired, learned by means of programmed learning, or adopted through plastic learning) and the context it is supposed to represent; and b) to lie to other members of one's own species. Both imply a conscious understanding of the cause-and-effect relationship involved and a capacity for self-direction.

Viewed in this way, the behavior of higher primates looks suspiciously similar to our own, but we find nothing in the literature or our own experience to indicate any awareness on the part of insects. Indeed, given the repeated successes of modern ethology's reductionistic "well-programmed robot" view to account in a predictive manner for ever-more-complex behavior, we fail to see how the hypothesis of animal consciousness has any major value at this stage for furthering our understanding of insect behavior (or, for that matter, most of the behavior of vertebrates). We think that for now, the course most likely to promote a deeper understanding of behavior is a fuller exploration of complex programming and innately directed learning so that we can better define how far beyond our present limits of understanding and imagination the principles of behavioral programming can take us before we must, out of explanatory desperation, invoke notions of consciousness to account for behavior.

Acknowledgment. Supported by grants from the Harry Frank Guggenheim Foundation and the National Science Foundation (BNS 78-24754).

REFERENCES

(1) Beer, C. 1960. Incubation and nest building by the blackheaded gull. Thesis: Oxford University.

(2) Boch, R. 1956. Die Tänze der Bienen bei nahen und fernen Trachtquellen. Z. vergl. Physiol. 38: 136-167.

(3) Boch, R. 1957. Rassenmäßige Unterschiede bei den Tänzen der Honigbiene. Z. vergl. Physiol. 40: 289-320.

(4) Brines, M.L., and Gould, J.L. 1979. Bees have rules. Science 206: 571-573.

(5) Bogdany, F.J. 1978. Linkage of learning signals in honey bee orientation. Behav. Ecol. Sociobio. 3: 323-336.

(6) Booker, R., and Quinn, W.G. 1981. Conditioning of leg position in normal and mutant Drosophila. Proc. Natl. Acad. Sci.: in press.

(7) Byers, D.; Davis, R.L.; and Kiger, J.A. 1981. A defect of cyclic AMP metabolism in the dunce mutation of Drosophila. Nature: in press.

(8) Cullen, E. 1957. Adaptations in the kittiwake to cliff nesting. Ibis 99: 275-302.

(9) Curio, E.; Ernst, V.; and Vieth, W. 1978. Cultural transmission of enemy recognition. Science 202: 899-901.

(10) Esch, H. 1964. Beiträge zum Problem Entfernungsweisung in den Schwänzeltanzen der Honigbienen. Z. vergl. Physiol. 48: 534-546.

(11) Fabre, J.H. 1915. The Hunting Wasps. New York: Dodd, Mead & Co.

(12) Farner, D.S., and Lewis, R.A. 1971. Photoperiodism and reproductive cycles in birds. Photophysiol. 6: 325-370.

(13) Frisch, K. v. 1946. Die Tänze der Bienen. Österr. Zool. Zeit. 1: 1-48.

(14) Frisch, K. v. 1967. Dance Language and Orientation of Bees. Cambridge, MA: Harvard University Press.

(15) Gallup, G.G. 1979. Self-awareness in primates. Am. Sci. 67: 417-421.

(16) Gardner, R.A., and Gardner, B.T. 1969. Teaching sign language to a chimpanzee. Science 165: 664-672.

(17) Gould, J.L. 1974. Genetics and molecular ethology. Z. Tierpsychol. 36: 267-292.

(18) Gould, J.L. 1975. Honey bee communication - the dance-language controversy. Thesis, Rockefeller University, New York.

(19) Gould, J.L. 1975. Honey bee recruitment. Science 189: 685-693.

(20) Gould, J.L. 1975. Communication of distance information by honey bees. J. Comp. Physiol. 104: 161-173.

(21) Gould, J.L. 1980. Sun compensation by bees. Science 207: 545-547.

(22) Gould, J.L., and Gould, C.G. 1981. The instinct to learn. Science 81 2(4): 44-50.

(23) Griffin, D.R. 1976. The Question of Animal Awareness. New York: Rockefeller University Press.

(24) Hertz, M. 1930. Die Organisation des optischen Feldes bei der Biene, II. Z. vergl. Physiol. 11: 107-145.

(25) Hinde, R.A., and Fisher, J. 1951. Further observations on the opening of milk bottles by birds. Brit. Birds 44: 393-396.

(26) Hölldobler, B.H. 1971. Communication between ants and their guests. Sci. Am. 224(3): 86-93.

(27) Janzen, D.H. 1974. Deflowering of Central America. Nat. Hist. 83: 48-53.

(28) Kandel, E. 1979. Small systems of neurons. Sci. Am. 241(3): 66-76.

(29) Kawai, M. 1965. Newly acquired pre-cultural behavior of a natural troop of Japanese monkeys. Primates 6: 1-30.

(30) Klinghammer, E., and Hess, E.H. 1964. Parental feeding in ring doves: innate or learned? Z. Tierpsychol. 21: 338-347.

(31) Lindauer, M. 1954. Temperaturregulierung und Wasserhaushalt im Bienenstaat. Z. vergl. Physiol. 36: 391-432.

(32) Lindauer, M. 1955. Schwarmbienen auf Wohnungssuche. Z. vergl. Physiol. 37: 263-324.

(33) Lindauer, M. 1963. Kompaßorientierung. Ergebn. Biol. 26: 158-181.

(34) Lorenz, K.Z., and Tinbergen, N. 1938. Taxis und Instinkthandlung in der Eirollbewegung der Graugans. Z. Tierpsychol. 2: 328-342.

(35) Lovell, H.B. 1963. Sources of nectar and pollen. In Hive and Honey Bee, ed. R.A. Grout, pp. 191-206. Hamilton, IL: Dadant.

(36) Marler, P. 1970. Song development in white-crowned sparrows. J. Comp. Physiol. Psychol. 71: 1-25.

(37) Marler, P. 1972. The drive to survive. In Marvels of Animal Behavior. Washington, D.C.: National Geographic Society.

(38) Marler, P., and Peters, S. 1977. Selective vocal learning in a sparrow. Science 198: 519-521.

(39) Menzel, R., and Erber, J. 1978. Learning and memory in bees. Sci. Am. 239(1): 102-110.

(40) Opfinger, E. 1931. Über die Orientierung der Biene an der Futterquelle. Z. vergl. Physiol. 15: 431-487.

(41) Pankiw, P. 1967. Studies of honey bees on alfalfa flowers. J. Apic. Res. 6: 105-112.

(42) Premack, D. 1971. Language in Chimpanzee? Science 182: 943-945.

(43) Quinn, W.G. 1976. Memory phases in Drosophila. Nature 262: 576-577.

(44) Reinhardt, J.F. 1952. Responses of honey bees to alfalfa flowers. Am. Nat. 86: 257-275.

(45) Savage-Rumbaugh, E.S.; Rumbaugh, D.M.; and Boysen, S. 1978. Symbolic communication between two chimpanzees. Science 201: 641-644.

(46) Seeley, T.D., and Morse, R.A. 1978. Nest site selection by the honey bee. Insectes Sociaux 25: 323-337.

(47) Seyfarth, R.M.; Cheney, D.L.; and Marler, P. 1980. Monkey responses to three different alarm calls. Science 210: 801-803.

(48) Smith, J.M. 1976. Evolution and the theory of games. Am. Sci. 64: 41-45.

(49) Spalding, D.A. 1873. Instinct. MacMillans Magazine 27: 282-293.

(50) Stent, G.S.; Kristan, W.B.; Friesen, W.O.; Ort, C.A.; Poon, M.; and Calabrese, L. 1978. Neuronal generation of the leech swimming movement. Science 200: 1348-1357.

(51) Tinbergen, N. 1960. The Herring Gull's World. Garden City, NY: Doubleday.

(52) Tinbergen, N.; Broekhuysen, G.J.; Feekes, F.; Houghton, J.C.W.; Kruuk, H.; and Szule, E. 1963. Egg shell removal by the black-headed gull. Behav. 19: 74-117.

Animal Mind - Human Mind, ed. D.R. Griffin, pp. 299-331.
Dahlem Konferenzen 1982. Berlin, Heidelberg, New York: Springer-Verlag.

Cognitive Aspects of Ape Language Experiments

C. A. Ristau and D. Robbins
The Rockefeller University
New York, NY 10021, USA

Abstract. Results from the ape language projects can be interpreted as revealing cognitive abilities. Some of the work begins to deal with the issue of meaning. Does the ape use individual lexical items in "word-like," conceptual ways, and can he combine the words in meaningful ways, utilizing some simple grammatical rules? The distinction is raised between the animal's "knowing how" to obtain a reward through instrumental learning and "knowing that" he is requesting a reward. An intensional analysis of meaning, a logical analysis describing the complexity of mental states, is also suggested as an alternative approach. Are the apes engaged in intentional (purposeful) communication? Experiments on deception begin to investigate that issue. Observations from field research may suggest fruitful improvements of ape language studies.

Recent attempts to teach captive apes something approximating human language have aroused great public and scientific interest. Limitations of space prevent a complete review in this paper, but extensive discussions and reviews have been published by Premack (26), Rumbaugh (30), Premack and Woodruff (27), Savage-Rumbaugh et al. (33), and Ristau and Robbins (28); elsewhere we will present an updated review (29). In this paper we will examine specifically those aspects of the recent experiments which appear to throw some light on the questions of mental experience and cognitive skills.

INTRODUCTION

Originally, attempts were made to teach apes vocal language (15, 17). Although the apes failed to produce more than a very few raspy utterances, researchers noted that the apes appeared to understand many spoken English words. More recently, three basic approaches have been used: (a) a manual gestural system, (b) a computer-based lexigram system, and (c) a problem solving approach utilizing plastic chips.

(a) As pioneered by the Gardners (9, 11, 12), manual gestures drawn from American Sign Language (ASL) were taught to captive chimpanzees. The initial project involved one chimpanzee, Washoe. More recently, the Gardners have used fluent human signers to teach several chimpanzees; exposure to ASL began at an age of only a few days. The Gardners' work with Washoe was continued by Fouts and extended to other chimpanzees; Fouts and his colleagues likewise taught a chimpanzee to transfer hand signs from English names to the objects they represented (8). Another chimpanzee, Nim, was taught hand signing by Terrace and his colleagues (38, 39). This work is particularly notable for the extensive data analysis of multi-sign utterances, including those drawn from videotape. While most experimental subjects have been chimpanzees, Patterson (24) has taught a gorilla Koko to use hand signs and to respond to spoken English commands, and Miles (personal communication) is teaching ASL to a captive orangutan.

(b) An artificial language, Yerkish, utilizing lexigrams (geometric diagrams) was developed as part of a computer system by Rumbaugh and his colleagues. The chimpanzee, Lana, was the first subject taught to use a keyboard with lexigrams to obtain, via the computer, various desired objects and events. She has learned to press appropriate sequences of keys to communicate, to some degree, with the computer and human experimenters; data from these interactions are stored in the computer.

(c) Premack and his colleagues (26) have developed a type of communication system in which trained chimpanzees use small colored plastic chips of various shapes which serve, at least to some extent, as symbols for objects and actions. In problem solving tasks, the chimpanzees are required to select chips or place them in appropriate sequences. Most data were obtained from his able pupil, Sarah, the chimpanzee.

Because the work with chimpanzees is the most extensive and, from published work, appears to be better controlled, almost all of the following discussion will involve this species.

SOME MAJOR ISSUES

Meaning

A central issue that has received experimental attention and vigorous discussion is the degree to which manual gestures or use of artificial lexicons has allowed chimpanzees to use something equivalent to single words and whether, in addition, some of the more proficient animals have learned to combine such word-like symbols in meaningful ways so as to generate something at least remotely comparable to the grammar of human languages. It is no simple matter to demonstrate either. For a sign or lexigram to be word-like or symbolic, it must reflect more than a mere association between that lexigram and some specific object; a word represents or "stands for" its referent in a way that an associated stimulus does not. Furthermore, we may well expect developmental differentiation of meaning, much as young children progress from initially using sounds and "words" holistically to using words both more specifically and more abstractly and combining them according to grammatical rules. Do apes?

"Knowing How" vs. "Knowing That"

Ryle (32) has distinguished between "knowing how" and "knowing that." As used by the philosophers, this distinction seems to be the central issue in various psychologists' criticisms that ape language usage is instrumental and related merely to the gaining of reward, i.e., the ape is demonstrating "knowing how" to get a reward but does not understand what he is "saying." In essence this distinction requires that we consider whether the signing ape or other communicating animal is consciously aware of the message it is conveying and of the results likely to follow. It has been customary in the behavioral sciences to ignore this possibility and to concentrate on input-output relationships and on patterns of information processing that must be postulated to explain the observed complex behavior. But to "know that" means to understand relationships beyond the coupling of a stimulus pattern to a response, no matter how simple or complex these may be.

The distinction between "knowing how" and "knowing that" seems intuitively accurate for certain kinds of descriptions. For instance, the accomplished pianist "knows how" to play a familiar piece, yet, typically, he is not, <u>at the time</u>, aware that his fingers are moving in a certain sequence pausing prescribed units of time. Later, he can "know that" his fingers were moving in particular ways by remembering his actions or, with minor movements of his fingers, replaying the piece. Also, if he so wishes, he can, while playing the concert, reflect upon his actions and simultaneously "know that" whilst "knowing how." Such reflection may, however, adversely affect "knowing how" to play, in that the skill of his performance may deteriorate.

We recognize that the distinction between "knowing how" and "knowing that" is not absolute; there are indeed penumbral cases as philosophers have noted (e.g., Dennett (5), p. 184). There are also different kinds of "knowing how" ranging from automatous walking upstairs to skilled mountain climbing. Arguments, beyond the scope of this paper, have been offered in which "knowing how" can be reduced to "knowing that" (Fodor, Dennett). Yet the distinction may nevertheless be useful in discussing the ape language research insofar as it illuminates the difference between rote performance to obtain reward vs. knowledge that entails awareness of the behavior <u>as</u> communicative and of the possible outcomes and ascribes meaning to the elements of the behavior.

Intensional Analysis

Another analysis that can be applied, perhaps more fruitfully, to the accomplishments of the apes and indeed to natural animal and human communication was described at this Dahlem Conference by Dennett (5); this structure is intensional analysis, deriving from analyses offered by Grice (14) and others (e.g., Bennett (2, 3)). "Intensionality" (spelled with an "s") is to be distinguished from the everyday term "intentionality," the latter meaning "on purpose." An intentional or unintentional act can be subjected to the logical "intensional" analysis which is a means of describing the degree of complexity of mental systems. A few examples of first, second, and third order intensional analyses follow:

1st order "A knows that p," where p is some proposition.
"A wants q."
2nd order "A believes that B believes that p."
3rd order "A wants B to believe that A wants x."

Such an analysis is useful in suggesting experiments, both in the laboratory and in the field, in reporting field observations including records of events that have been observed only once (anecdotes), in sensitizing the field researcher to pertinent naturally occurring events, and in examining the ape language experiments that have been conducted in order to better understand the level of mental complexity implied by each class of behaviors. These issues are discussed further in the state of the art report in this volume on "Communication as Evidence of Thinking" (Seyfarth et al.).

INVESTIGATIONS INTO MEANING

At least the following kinds of evidence have been offered as relevant to meaning for a chimpanzee: (a) labeling of an object, (b) novel uses of a word, (c) errors made in generalization, (d) novel word combinations to refer to previously unlabeled objects or situations, (e) feature analysis of an object and its label (26), (f) functional definitions of words (33), and (g) categorical sorting (35). As we consider the significance of these kinds of evidence to meaning, we must realize that the definition of meaning is a murky issue, unresolved by philosophers and often not considered in its complexities by the ape language experimenters.

Labeling

Training to name has been an integral part of all the ape projects. In the signing projects, labeling was deliberately and routinely taught and tested by "This-X" statements and "What-this?" questions. Observational learning for names was formally attempted by Premack using plastic chips; but it met with little success.

The distinction between the apes' "knowing how" vs. "knowing that" is raised by some of the results from the original Lana project. Lana routinely and correctly used phrases of lexigrams to "request" specific rewards, but she apparently did not know the meaning of individual lexigrams in her production as indicated by her performance with the lexigrams in a labeling task. One could describe her behavior as "knowing how" to obtain a reward, but not "knowing that" she was requesting a reward.

Alternatively one might use an intensional analysis. When Lana presses the lexigram keys "Please machine give banana," is her behavior best described as "Lana wants a banana" or as "Lana wants the trainer, Tim (or the computer?) to give her a banana" (both first order) or as "Lana wants the trainer to believe that she wants a banana" (third order)? Observing Lana's spontaneous, non-trained behavior using the lexigrams in a variety of novel contexts and devising more experiments could allow us to determine which is the best level of description. It is beyond the scope of this paper to apply an intensional analysis to the ape language research that will be discussed, but, as suggested by Dennett, the idea seems a very fruitful one.

Note that Lana required 1600 trials to learn the names of two items, banana and M&M (candy). Previously she had requested these items hundreds of times using such stock sequences as "Please machine give M&M." She was then taught to name by a paradigm in which she was presented with either a banana or an M&M on a tray, and was asked the question on the experimenter's keyboard "?What name-of this." She was to respond via the keyboard, "M&M name of this." Correct responses were rewarded with the object or amusement of her choice (31). The fact that Lana required so many trials to learn this task suggests two possible interpretations: a) it took that long for Lana to learn the nature of the naming task, and/or b) Lana did not know the meaning of the individual lexical items when she had previously requested M&M and other products. Subsequent training with other items resulted in much more rapid labeling, suggesting the initial difficulties in the labeling task lay in distinguishing the requirements of that paradigm from the task in the requesting paradigm in which the lexigrams had previously been used. However, it is also clear that as the work was originally described by the researchers (30, 31), Lana's use of the stock sequences ("knowing how") was considered to imply understanding of the individual lexical terms (part of "knowing that"). In subsequent work, the researchers have sought to investigate more fully what such "understanding" might entail.

Premack's work also necessarily dealt with labeling exemplars. The typical way of naming an item, e.g., figs, was to introduce them into a situation of the kind already "mapped" or well-known to Sarah, such as giving. Then she would be induced to use "fig" in a sequence, the other "words" of which she knew. For example, the trainer, Mary, would place a fig before Sarah, give Sarah a pile of four plastic chips, namely, "Mary,""Sarah," "fig," and "give" and induce her to write the sequence, "Mary give Sarah fig." Premack interprets her behavior as learning the name for "fig" (26). In the least, Sarah could substitute the lexigram for "fig" appropriately in a variety of sequences.

The concept "name of" was taught by means of two positive and two negative exemplars. For example, the plastic chip for "apple" and an actual apple were placed with a gap between them, and Sarah was given only one plastic chip, the new one intended to mean "name of." By putting that plastic chip between the two items, she created a sequence, intended to be read as: "'Apple' name-of object apple" (26). "Not-name of" was formed by gluing the negative particle to the plastic chip for "name-of." Then the plastic chip for "apple" and the object banana were placed slightly apart from each other, with "not-name of" the only chip available to place between them. Then the terms for "name-of" and "not-name of" were used to introduce the plastic chips intended to represent new items. The chimpanzees, irrespective of training technique, were able to use the labels correctly both in tests of naming and in sequences with other plastic chips.

We can wonder if the chimpanzees were learning the "names" of items and were forming the concepts "name-of" and "not-name of." This is a very general problem of determining the abilities underlying specific behaviors of an organism or particular aspects of a computer program attempting to model artificial intelligence. McDermott (19) terms the tendency to overinterpret specific and limited capacities "wishful mnemonics." One should instead label a capacity neutrally, e.g., call it "6PX4" and then, by virtue of the actual behaviors of the organism, slowly build a notion of what "6PX4" might entail. The point is well taken.

In the case of "naming" for instance, we might better attempt to specify the attributes of "naming" and suggest the necessary experimental evidence to establish the occurrence of naming. As we use the term in everyday language, a "name" is a symbol. A symbol is many things. In part the symbolic use of a word involves "displacement," the ability to speak of objects or events remote in space and time. It also involves freedom from task and context specificity so that the symbol can be used cognitively in a variety of ways. Linguists and philosophers have suggested still other characteristics which are beyond the scope of this paper. Some aspects of the ape's symbolic use of a lexigram could be revealed by the following procedure suggested by Terrace (personal communication): one could investigate the ape's ability to respond to questions concerning the referents of the lexigrams even when the lexigrams are not paired with the referent food items. Clearly the "labeling" paradigm as it is used by all of the ape language researchers does not include such questions.

Yet other paradigms to be considered in this paper (e.g., Premack's feature analysis of an object and its label and the categorical sorting task of Savage-Rumbaugh et al.) and possible spontaneous uses of labels do include instances in which lexigrams are used to refer to objects not immediately present. Such paradigms do reveal symbolic aspects of at least some chimpanzee productions. Labeling training is a possible first step in the process of word or concept or symbol acquisition. Labeling in the ape language work is by no means equivalent to a symbol or to the richness of a human word.

The various signing projects used Ameslan signs as labels for objects and events. However, the hand signs have been criticized for their iconicity. Indeed, a number of Ameslan signs are iconic, that is, the sign is similar in form to or imitates the action or object which is signed about. Savage-Rumbaugh et al. (33) have claimed that apes' use of iconic signs does not reveal referential ability as the use of more arbitrary signals would, but rather draws on the apes' abilities to note perceptual similarities. Yet if we consider the so-called obvious iconicity of the sign for tree, which entails an upright forearm (the tree trunk) with fingers spread wide (the tree tops), the iconic aspects are obvious, only after one is made aware of the meaning of that sign. In fact, tests given to nonsigning humans on the meaning of signs, reveal that such persons do very poorly (1). Once the naive observers are told the meanings, however, they generally agree on the representational basis. Even if the chimpanzee is more adroit than humans at guessing the iconic referent for some sign (which is unlikely), one must keep two points in mind: a) Not all or even most ASL signs are icons, and b) it is an impressive intellectual feat to comprehend the abstract relationship that five open fingers represent the many leaves and limbs of a tree. Rather than being a denigration of the apes' abilities, a capacity to appreciate the iconicity of signs, if the apes have such an ability, should be considered instead an exciting cognitive skill to be investigated further.

In summary, although there are various methodological and interpretive problems associated with each project, within each project, the chimpanzees, using some labels, have been able to request objects, even in the absence of the object itself, respond appropriately in naming tests, and use labels correctly in problem-solving tasks. Yet, as we have briefly discussed, labeling alone is not all we mean by the symbolic nature of a word.

Novel Use of a Word

The issue of novel word use becomes closely enmeshed with the issue of word generalization and overgeneralization. Designation of "appropriate novel use" often depends upon the chimpanzees' abstracting features of the environment we consider prominent and using a word we consider pertinent. To deal with the issue of novelty and its implications for meaning of a label, we need to have adequate sampling and quantitative measures of occurrence of labels and the environmental contexts of their use.

The Gardners do mention examples of generalization although neither they, nor anyone else, give quantitative measures of use. For instance, they note that "more" was generalized and used "for continuation or repetition of activities," and for additional portions of food, and other items. It was often combined as in "more go" and "more fruit" (9).

Errors Made in Generalization

The way a sign is initially generalized can reveal the tremendous difficulty inherent in determining the chimpanzees' meaning for a sign that we may gloss quite differently with respect to referent and semantic class, i.e., whether the sign functions as a verb, noun, preposition, or nothing of the sort to the ape. As an example, Washoe was introduced to the sign for flower by being presented with a real flower. However, she apparently interpreted this sign to be associated with smell, so she generalized it to pipe tobacco and kitchen fumes (12). Since the sign for flower entails holding the fingertips below the nose and breathing in, an act similar to smelling the fingers, generalizing the sign to odor is not unlikely. Apparently the trainers often drew Washoe's attention to odor by breathing in deeply and obviously smelling (Savage-Rumbaugh, personal communication). Yet the fact remains, there are many features of the situation that could have been selected as outstanding to the ape and used when employing the sign in new contexts; the ape selected odor or the act of smelling. Her behavior suggests one might attempt to present apes with one or few instances for a sign and then note the natural generalizations the apes make, thereby suggesting to us what the most salient characteristics of the situation are to the ape mind.

Novel Word Combinations

The anecdotes that report Koko calling a face mask an "eye hat" and Washoe calling a swan a "water-bird" are intriguing, and potentially illuminating concerning ape mentality. The possible selective reporting of data by investigators and the lack of carefully reported anecdotes and statistical measures of the occurrence of those novel combinations and of possibly other nonsensical combinations make one somewhat restrained in enthusiasm. Yet the data are potentially very important.

For instance, it has been noted that Nim often combined one of his favorite foods in topics of conversation, the "banana," with a huge variety of words - sorry, drink, tickle, toothbrush, hat, and hand cream. It seems unlikely that on each occasion he was trying to say something profound. Yet as Desmond suggests (6), it is difficult to dismiss preposterous combinations such as "banana toothbrush" which do occur, for there are plausible communications the chimpanzee might be attempting to make with that combination. He might, for instance, be requesting a banana to eat and a toothbrush to clean his teeth (6).

Some of these combinations might be word play, an important characteristic of children's language; apes are sometimes slighted for apparent lack of such play. Yet for whatever its significance, Washoe and other apes have been reported "signing to themselves" when no one else was present. One might in fact be able to distinguish between word play and novel use of words by using an ethological approach; perhaps one could note the facial expressions and other behaviors that accompany such uses and then relate these to behaviors common in play.

Terrace (personal communication) notes that it is implausible for Nim to be requesting simultaneously a banana and a toothbrush for both were unlikely to be in view at the same time and Nim almost never signed for objects that were not physically in view. The latter fact, if applicable to all the signing apes, is an extremely important kind of information about their use of artificial language for it implies that the chimpanzees very rarely exhibit displacement, a criterion for a symbol. Clearly, however, there are at least limited circumstances in which apes can be required to exhibit displacement, e.g., Premack's feature analysis, and the tool use and exchange paradigm of Savage-Rumbaugh et al.

Feature Analysis of an Object and Its Label

Premack was interested in asking whether a chimpanzee is able to analyze an object into its features and whether a match-to-sample technique could be used to answer that question (26).

In the chimpanzee Sarah's first such test, the sample she saw was an actual apple. She was to "match" to this sample by choosing from pairs of alternatives that did and did not instance some feature of apple. None of her choices were rewarded in preference to others and all of her choices were acknowledged with approving tones. In order to maintain her interest in the task, she was occasionally allowed a choice of food rewards. Premack notes that her feature analysis of the apple accords well with human analysis. She chose red over green, a square with a stemlike projection over a plain square, a plain square over a circle, but was inconsistent in her choice of roundness (seven choices out of ten) over a square with stem (three choices out of ten). The test was then repeated except that the object apple was replaced with the word for "apple," namely, a small blue triangular piece of plastic. Her choices for this task were essentially the same, even making the same seven to three split on round vs. square with stem.

Premack has been criticized for interpreting these results to mean that Sarah had a similar concept for the word "apple" as she did for the object apple (34). This criticism might be justified, for chimpanzees have good memories, and Sarah could well be merely repeating the responses she made for the object apple when she was presented with the word "apple." However, because Premack did the following experiment, these criticisms are not relevant. The critical experiment was to reverse the order of the task, that is, present choices for the word for an object first and then present choices for the object itself. In such a test, Sarah was shown the word for a caramel and then was given four pairs of alternatives that were intended to instance the shape, color, size, and texture of a caramel, a brown cube of candy wrapped in cellophane. Again the features she chose generally accorded with human analysis and were approximately the same for the word and the object.

Finally she was given a feature analysis of the word "apple," using words rather than objects as alternatives when possible. Her agreement between the analysis of the word and the analysis of the object was about 96%. In a different variation Sarah was asked to describe the attributes of the "word" for apple and was given alternatives irrelevant to the object apple and relevant to the little blue plastic triangle, the name of apple. In this experiment, unlike the previous ones, Sarah's incorrect choices were specifically disapproved of. Her alternatives were objects and plastic chip names of attributes of the blue triangle, such as "round" vs. "triangle." She erred once in the 15 pairs of alternatives. Premack interprets Sarah's abilities as comparable to using a dictionary in both directions. Given the plastic chip name for apple, she can select the alternative pertinent to the object apple, and given the object, she can do the same for the word. Since both descriptions were made in the absence of the item that was described, "a need for memory or internal representation" was introduced (26). This general experimental approach is an interesting one to develop more broadly, analyzing more objects and concepts than apples and caramels and offering the ape a greater variety of alternatives to choose among.

Functional Definitions of Words

Savage-Rumbaugh et al. (33) report experiments in which two chimpanzees learned to use lexigrams to communicate to each other about specific tools needed to obtain food which they then shared. The experiments are too complicated to describe in detail and the specific issues addressed by the experiments are open to debate and beyond the scope of this paper. Among the topics of importance are a) the process by which a lexigram may come to represent objects and events and thereby begin to acquire symbolic functions, and b) the development of turn-taking and productive and receptive communicatory functions between two apes (see (29) for a fuller discussion).

From the reviewers' point of view at least two important points have emerged that deal with the process of word acquisition. One is that Savage-Rumbaugh et al. have noted that chimpanzees were predisposed at first to associate the lexigram intended to represent a specific tool with a location rather than with a tool. Later, with the introduction of additional tool sites for each tool, the chimpanzees did learn to associate the lexigram with a specific tool. Second, there was a very great difference in the way in which chimpanzees learned to label a tool according to the typical naming paradigm and the way they learned to request a tool when needed or to select the correct tool when it was requested. The labeling process took a very long time, namely, hundreds of trials, while the functional tool use method required tens of trials. The distinction between functional vs. labeling training is potentially a very important one. Functional training relates to ideas of Piaget and Bruner according to which a child is said to first learn the meaning of a word in terms of the child's action which is associated with the word. Thus the meaning of the word "hole" is "to dig." Active manipulation is also critically important in perceptual development, specifically in limb-eye coordination (17). Early childhood educational curricula emphasize the importance of learning mathematical and spatial concepts through active manipulation of objects. The Gardners' recent signing project also has findings relevant to this issue. From the array of referents presented to young children in a study by Nelson and to the Gardners' "linguistic" apes, those commonly named were "changeable, movable, and manipulable" (12, 23). For example, household items such as sofas and tables were not named, but lights which children switch on and off were named. However, in the research of Savage-Rumbaugh et al., it is not clear precisely which variables are important for the great differences in learning between functional and labeling training. It is possible that the greater number of distinctive cues associated with functional training are the main source of the differences. Further research should investigate these matters.

Categorical Sorting

Even in the early ape language attempts (15, 17), researchers noted that the apes could sort items, including pictures. Recently, Savage-Rumbaugh and colleagues (35) have utilized a categorical sorting task to explore whether lexigrams used by chimpanzees were functioning at a representational (or symbolic) level or whether the lexigrams were merely paired-associates with the items they "labeled." To answer this question, the chimpanzees were trained to label the names of three inedibles as tools and the names of three edibles as foods. The chimpanzees' ability to generalize these categories was tested by presenting them with the names of other foods and tools and asking them to label these additional names as foods or tools. The essential point is that the apes categorized the names of objects, in the absence of the objects themselves. To do this, the researchers state, the apes must have an internal representation of each object and of the characteristics it shares with the category. Note also that Lana, long trained to "label" objects could not do this categorization task, while two other chimpanzees, taught by "functional" training methods could.

The results may reflect Lana's overtraining in the "labeling" experimental paradigm, rather than a lack of categorical understanding of the labels. Some arguments may be raised as well about the exact nature of the other apes' cognitive abilities. (further discussion in (29)). The approach is an interesting one and should be extended to other categories and to a variety of contexts in which names for objects are learned.

EVIDENCE FOR GRAMMAR

Meaning can accrue from the order of lexical items within an utterance. The ape language projects offer several approaches for investigating rules that might govern such order. Some are: a) the ability to make lexical substitutions was explored by training chimpanzees to produce sequential orders of plastic chip tokens (26); b) the ordering used by a signing chimpanzee, Washoe, was studied by a linguist (20); and c) records of chimpanzee signing including videotape or transcripts were analyzed for lexical regularities and were compared to the specific prior utterances of trainers to determine initiative responses and spontaneity (39).

Lexical Substitutions

In Premack's work, the chimpanzees were regularly asked to produce sequences such as "Mary give Sarah fig" after observing the behavioral sequence or else to substitute into gaps in the sequencing one or more plastic chips from a small set of chips available to the chimpanzee. Appropriate selection and placement of chips were considered to be evidence for understanding lexical categories.

This work, and the Lana project as well, has been criticized for using extensive multiple-choice tests which may merely require rote memory. The Gardners (11) argue that previous experimental laboratory work (7) has shown that chimpanzees can learn to make the correct choice for each of 24 different arrays of objects, and that this capability could account for much of the performance exhibited by Premack's Sarah, as well as Lana (30). In addition, the Gardners (13) point out that test sophisticated rhesus monkeys show Harlow's learning set phenomenon, namely, a large increase in the improvement between trial 1 and trial 2 as the monkeys solved more and more problems. In some cases performance increased from a chance level on trial 1 to almost 100% correct on trial 2. Therefore, they argue, only in trial 1 of transfer tests are we measuring transfer. Analysis of trial 1 behavior would answer these objections, but trial 1 data of Sarah's and Lana's performance are seldom presented.

Terrace (37) argues that much of Premack's work on lexical understanding is problem solving. He notes that tests of the use of prepositions never contrasted prepositions with one another, but rather had the chimpanzee demonstrate objective knowledge of a specific preposition. For example, to test the apes' understanding of "on," Premack required Sarah to identify which object was on top. That is, she was presented with the plastic chips representing "?red on green" or "?green on red" and was required to answer "yes" or "no." Furthermore Terrace states that, with few exceptions, most of the problems presented during training sessions were of the same nature and, within a session, the number of alternatives were limited. Terrace raises the query, "Is Sarah showing discrimination learning rather than language?" There is a brief report available of work by Lenneberg (18) indicating that high school students performed very well on a set of problems similar to those given by Premack to Sarah. The students, however, did not realize, despite their able performance, that the chips were supposed to represent "words" or that the sequences of chips were supposed to be linguistic. Yet the brevity of this posthumous report and the lack of details of experimental methodology lessens the impact of the results.

Linguistic Analysis of Washoe's Signing

As analyzed by McNeill (20), the order of signs in Washoe's utterances tended to reflect "addressee-nonaddressee," that is, the individual to which communication is addressed followed by any other individual or object. In the short, simple word combinations used by the signing apes, such sequencing by the ape accords well with our structural analysis of "agent-object." An example might be an ape signing "Mary give banana." Apart from addressee and nonaddressee, McNeill further suggests, word order probably depends on the prominence of the things referred to. He also suggests, as have Terrace et al. (39), that words were often added to the strings not to convey new information, but to make the message of the string more emphatic and intense. McNeill gives as an example, "Please open hurry" (20).

McNeill likewise recognizes that social hierarchy is important in Washoe's signing. He notes that Washoe's signing yields many examples of appeasement and begging to the human trainers. For instance, in the signs for "Please come-gimme," the ASL sign for "come gimme," at least as used in the Washoe project, is identical to the natural chimpanzee begging gesture (20). McNeill further suggests that typical messages of the chimpanzee will probably consist of "entreaties, demands, mollifications and declarations of ownership and location" (20). These suggestions find a parallel in research into natural animal communication systems. There are some difficulties with this approach, for it is not clear that a quantitative analysis has been done. Furthermore, in the Washoe project, the actual order of the ape's signing was not always precisely recorded in the written record; for instance, "tickle me" and "me tickle" were considered to be equivalent and might be recorded in either order.

Analyses to Determine Lexical Regularities and Imitation, Spontaneity and Turn-taking by Signing Chimpanzees

The signing project by Terrace and his colleagues (39) is outstanding primarily because of the extensive analyses done of Nim's multi-sign utterances; more than 19,000 of Nim's combinations were studied in some analyses and three and a half hours of videotapes were investigated in others. The data were analyzed initially to determine whether Nim's productions exhibited any lexical regularities. To answer this question, a large sample of the chimpanzee's utterances was analyzed for lexical regularity. Independent position habits did not account for the order in two sign utterances; there were lexical regularities in Nim's two sign utterances. However, there were not lexical regularities for three and four sign utterances, and there were insufficient data to analyze utterances of five signs or more.

A very important point about Nim's longer utterances is that, unlike longer utterances spoken by human children, Nim's did not add new information; they were redundant and repetitious. McNeill (20) had noted that Washoe's utterances tended to emphasize through repetition. For example, in the Nim project, the most frequent two sign utterances were "Play me," "Me Nim," "Tickle me," and "Eat Nim." The most frequent three sign utterances were "Play me Nim," "Eat me Nim," "Eat Nim eat," and "Tickle me Nim." The most frequent four sign utterances were "Eat drink eat drink," and "Eat Nim eat Nim" (39). A sixteen sign utterance made by Nim when the average length of his utterances was 1.6 was "Give orange me give eat orange me eat orange give me eat orange give me you" (39). Such repetitive sequences are not restricted to Nim. "Please milk please me like drink apple bottle" was signed by Koko when her average mean length of utterance was 1.75 (25).

We can also ask whether regularities that do exist are meaningful. To answer this question, a semantic analysis must be made for a wide variety of contexts in which the general environmental and social context is accounted for. In some analyses, Terrace uses the term context in a much more restricted way to refer to the prior signing made by humans. Without a broader definition of context, which is closer to ethological use of the term, one can miss regularities that do exist. For instance, "Me play" and "Play me" might conceivably be used equally often. One might interpret this to mean the order was a chance event, unless one had the information to know whether context determined which word order was used.

Terrace et al. did do a broader semantic analysis on data from videotape. It is clear videotaped data can have advantages over written or audiotaped data. A knotty problem remains in all the ape language work, in that videotaping is typically done under restricted conditions and times for photographic and economic reasons. Yet a study of "language" development requires data from a wide variety of contexts, and particularly in novel situations.

Terrace and his associates also performed a "discourse analysis" of videotape transcripts of Nim and his teacher signing in order to understand how Nim's sign sequences were related to those of the teacher's. The three and a half hours of videotapes were drawn from nine sessions most of which were training/drill sessions but also included some home sessions. Results indicated that a very large proportion of Nim's utterances were adjacent, defined as those that follow the trainer's utterance without a definitive pause, 87% for Nim as contrasted to a child of about 21 months old in the first stage of linguistic development who has 69% adjacent utterances. Of Nim's adjacent utterances, about 44% were imitations or slight reductions of the utterance just produced by the human trainer in contrast to 18% for children. Only 7% of Nim's utterances were expansions of the trainer's prior utterance in contrast to 21%.

Nim also did not appear to exhibit turn-taking as children do in signing or verbal conversation with an adult. Nim, for instance, signs simultaneously with the trainer 71% of the time, typically interrupting the trainer's signing.

Several criticisms have been levied against Terrace's work. 1) We noted that Terrace and his colleagues did do a semantic analysis accounting for social and environmental context, but contexts were limited to the conditions for which videotape data were available. 2) Nim's lack of turn-taking is claimed not to be as interruptive as stated by Terrace and his colleagues. In signing, just as in nonverbal communication, two people may sign or nod in agreement simultaneously without being considered as interrupting each other and without interfering with communication in the way two simultaneous spoken phrases could. 3) Nim experienced a very large number of trainers, 60, more than apes in any other language project. It must be noted that Nim did have some stable social emotional relationships, for at least five or six trainers were involved in the project for at least a year. Nim also had a core group of 8 teachers, while most of the other trainers were occasional playmates. Furthermore, chimpanzees in other projects did experience variety in trainers, e.g., the Gardners (11) report that chimpanzees in their recent project were regularly exposed to at least six different humans on any one day. Yet even Terrace agrees (39) that a stable and small group of caregivers should be used in any project that might attempt to surpass the limits of Nim's achievements.

4) The Nim project also appears to differ in its greater emphasis on structured training sessions or drills although it is very difficult to evaluate this matter adequately for precise data are at least difficult to find for other projects. Likewise, Nim's caretakers did try to teach him to sign during all of his waking hours, which included trips to a "gym," a park, and outdoor walks. The issue of drill is potentially an important one, for drilling <u>could</u> limit the spontaneity, variety, and creativity of an ape's productions both within and outside of the

training sessions. Miles (22) notes of her work with two chimpanzees that the most interesting and unusual sign language exchanges between a chimpanzee and a human occurred in the casual interaction between the apes and human trainers rather than during the experimental sessions. Thus the time before and after sessions, during walks in the woods, car rides, feeding, and general play tended to result in more spontaneous communication by the chimpanzees than the formal sessions did. Yet typically data, especially filmed records, are not available for such casual signing.

In summary, there is no convincing evidence for grammatical structure in the ape language projects. McNeill offers an insightful analysis of Washoe's signing as reflective of her social relationships and matters of prominence to her. Because of difficulties in the original Washoe project, including the use of nonfluent ASL signers and data collection methods that did not emphasize accurate report of sign order, that analysis should be made for the more recent, modified Gardners' project with four young chimpanzees.

DECEPTION: AN INTENTIONAL ACT OR ANOTHER WAY TO GET A REWARD?

We have so far considered meaning as it refers to individual lexical items. Yet developmentally words are considered to derive their meaning from the whole speech act which has meaning. Meaning refers to entire utterance or series of interactions. An interactant's intention becomes essential in determining meaning.

Woodruff and Premack (41) have undertaken a series of experiments to investigate intentional communication by the chimpanzee by exploring the chimpanzees' ability to deceive and to resist being deceived. In these experiments a human and a chimpanzee communicated about the location of hidden food. No artificial lexigrams were used; instead the human and the chimpanzee developed and used nonverbal signals. Each member of the human-chimpanzee dyad served alternately as "sender" and "receiver" of the information. When the human cooperated with the chimpanzee and gave all the hidden food, if found, to the chimpanzee, the chimpanzee was successful both at producing and comprehending behavioral signals about the food's location. When the human and chimpanzee competed for the goal and the human kept all the food for himself, the chimpanzee learned to withold behavioral information, to mislead the competitive receiver, and to discount the sender's misleading behavioral cues.

The fact that the chimpanzee can perform both the role of sender and recipient of misleading information is critical to the notion of intentionality. As Woodruff and Premack note, an assumption of truth is an important component of any conversation (14). When this assumption is violated, intentionality may be revealed by the interactant's ability to suppress information or otherwise adjust his responses to information. It is this voluntary aspect of the behavioral adjustment, dependent upon context, that is central to an assumption of "intentionality." There are field examples of deceptive behavior in a variety of species, for instance, the familiar one of certain birds feigning injury when offspring are in the presence of a predator. Critics claim the behavior is unintentional and "triggered" by the predator's presence. Yet the behavior is dependent on context only in a general sense; it is not always performed when a predator is near. Primates as well have been reported to exhibit apparently deceptive behavior (21, 40).

In the course of the Woodruff and Premack studies, a very interesting behavior emerged, that of the spontaneous development of pointing by the chimpanzees. The emergence of pointing with an outstretched arm or leg is all the more interesting, because in the lab or field situation, chimpanzees have rarely or never been observed to point; but chimpanzees do comprehend pointing by humans. As Woodruff and Premack note, the emergence of pointing is not easily explained in the traditional terms of the experimental psychologists. Pointing does not result from shaping or guidance nor does it depend upon differential reward.

It is not explicable in terms of least effort. Because the chimpanzees in the Woodruff and Premack studies had seen humans pointing, the development of the apes pointing could be the result of observational learning. If so, it is nevertheless true that pointing was not necessary to obtain food. By the time pointing emerged, the chimpanzee had been successful at obtaining food merely through the trainer seeing him glance in the appropriate direction. Woodruff and Premack suggest that pointing emerged in the chimpanzees, because it is easier to control a limb than it is to control cueing through eye gaze.

A series of photographs illustrates a chimpanzee, Sadie, lying to a selfish trainer. First the chimpanzee extends a limb towards the unbaited container while making eye contact with the competitive trainer, and then she looks and points at the unbaited container. Finally, after the trainer lifts the container and discovers no food, "Sadie's head snaps in the direction of the other container, which she knew contained the food" (27).

Yet we must ask whether the chimpanzees are indeed practicing deception or whether they have simply learned effective means of obtaining food in different situations. At least two facts suggest to us that the chimpanzees know they are being deceptive: a) new voluntary behavior, pointing, emerged and was used advantageously in two different contexts, to point to the correct or incorrect container; and b) the sequence of actions illustrated in the photographs of the lying chimpanzee, particularly when the chimpanzee's head "snapped" back to the baited container, suggests a knowing deception.

Firmer evidence that the chimpanzees are knowingly deceptive would result if they could be shown to form a concept of "truth" vs. "deception or falsehood." This might be demonstrated by giving the chimpanzees a sorting task with markers or plastic chips to represent "truth" and "lie." The chimpanzees might then be presented with a series of videotape sequences each followed by a photograph or another sequence representing an actor communicating an obvious falsehood or a veridical statement. The chimpanzee would have to mark each conclusion with the appropriate token for "truth" or "falsehood." Chimpanzees' accomplishments in such tasks might well be prone to the vagaries of interpretation that beset other pioneering work by Premack and Woodruff using videotape sequences to investigate the attribution of mental states by chimpanzees (27). We must also realize that deception is a communicative act requiring a high level of intensionality (5), and, therefore, sorting tasks alone are not adequate to provide evidence of deceptive ability; we would need a variety of lines of evidence establishing knowledge of truth and deceit by communicator and recipient.

CONSCIOUSNESS

Finally, the problem of consciousness continues to arise from various facets of the ape language research. For example, as the term intentional is commonly used with respect to language, it involves the notion of conscious purposefulness. An animal may be aware of objects and events in the external world, or of mental events, or it may also be conscious in the sense of being self-reflexive. That is, an animal could be aware that it is he who is communicating or seeing a certain color or intending to ask for a tool.

Data relevant to the notion of awareness of external objects may be found in Premack's work investigating Sarah's concept of "apple." That work differs from a concept formation task per se in that Sarah was asked to reveal, in a limited way, what it is about the concept of "appleness" that is important to her. The deception experiments, and more complex experiments on deception that could be devised, are also beginning to deal with the notion of conscious intention.

Possible examples of self description may also reveal a sense of self awareness which is distinguishable from mere awareness of external objects. For instance, one of Premack's chimpanzees, Elizabeth, on several occasions spontaneously described what she was doing. She was cutting an artificial apple with a knife and she displayed chips representing "Elizabeth apple cut." This and others are intriguing examples, though we recognize they are vulnerable to the same flaws of interpretation that have been ascribed to other facets of the ape language projects and which we have criticized. Yet we should not ignore these examples. We might, for instance, be able to devise a multiple-choice personality type questionnaire containing descriptive, "if-then" statements or other statements requiring causal analysis. In these various statements, the individual chimpanzee could be represented by a symbol. She could then be required to choose from a variety of alternatives which describe actual and nonsensical or impossible daily activities, what her feelings are, her preferences, and so forth. Other variants of such a task might include questions about herself as contrasted to questions about other chimpanzees. She might also be required to sort still or motion pictures of herself and of other individuals and to label them with the individual's name.

SUGGESTIONS FOR FUTURE RESEARCH

It may be helpful to look to the natural communication behavior of apes when developing experiments to analyze their cognitive abilities, possible precursors to language abilities, and perhaps even their mental states. Most of the words used in ape language projects are names of physical objects or descriptions of them such as color or size. Yet we know that young apes do far less manipulating of objects in their play than do humans; they pay more attention to social relationships. We might learn more about the apes' linguistic abilities if we attempt to teach them words describing social relationships and attributes such as "boss," "mother," "my offspring," "male," "female," "juvenile," and words to describe deference, neutrality, and dominance. We might also wish to concentrate on natural signals that are not highly arousing and thereby disruptive to our research and try to interpret the meaning of those signals. To accompany our good guesses, we might next attempt to teach signs which "map" those meanings. Thus we may possibly bridge the transition from ape calls and gestures (traditionally assumed to be largely instinctive) to more clearly voluntarily used signs or artificial representations. Children may begin to use language in this way and similar stages may also have been involved in the evolution of language by early man.

We might also search for "schema" exhibited by young apes in their play, perhaps similar to those of one to three year old children. Children, for instance, spend long times sorting items into categories that are obvious and not-so-obvious. At other times children put items <u>into</u> containers, remove them, put them back in again and again. One could speculate that children are laying the cognitive foundations for categorical analysis and for the preposition "in" and related grammatical structures. Do young apes exhibit similar behaviors?

Our field observations should also be directed to the apes' and other animals' cognitive abilities, some of which may already have been revealed in the laboratory. For instance, the recent work by Seyfarth, Cheney, and Marler (36) offers evidence that vervet monkeys may use predator alarm calls semantically, that is, the alarms represent different classes of predators. Yet we must be aware that the apes' burgeoning skills have been carefully guided in the laboratory, that their attention has been focussed on elements essential to the cognitive task, and that such guidance most probably does not exist in the field. The apes' natural abilities may appear less well developed. It is, perhaps, in the apes' complex social interactions that one may find most evidence of their cognitive skills.

We might look carefully to the <u>errors</u> that chimpanzees make, both in their generalization of signs to new objects or events and to their errors on concept formation tasks. We might also look closely at the different rates of learning various concepts. These may help reveal underlying proclivities of the ape mind.

Developing Premack's work, we could continue devising methods of teaching concepts that may underly language or other cognitive abilities. One such concept that has not yet been adequately demonstrated is the apes' understanding of time or temporal concepts. An animal's episodic memory could be analyzed by the use of markers for past, present, and future and by two indices for an event. One of these indices would be used by the experimenter to question whether the event indicated by the index occurred in the past, is currently underway, or will occur in the future. One of the indices could be one light of a sequence of colored lights while the other index could be the kind of reward received by the animal. The colored lights could be an invariant sequence, while the order of rewards changed. The experimenter could question the ape about an event designated by a colored light; the ape could verify the event in question by indicating the token representing the reward he received at that time (or none if it is a future event), and then the ape could indicate by a marker whether the event was past, occurring now, or future. Zaidel (42) has used a related, nonverbal paradigm in studies of right hemispheric functioning of split-brain humans.

The time may be ripe to stop avoiding the use of anecdotes, empathizing with animals and the topic of consciousness. We cannot escape from consciousness as the positivists and strict behaviorists have been unsuccessfully trying to do. Let us therefore strive to understand it. Let us consider the possibility that apes and other animals might think, and let us use past experimental research, field observations, our everyday intuitions, and guidance from philosophical analyses of such issues as communication acts, concepts of meaning and intentionality, to help us devise sound experimental, scientific approaches to the study of animal minds.

Acknowledgements. Supported by the Harry Frank Guggenheim Foundation. We would like to thank D.C. Dennett, D.R. Griffin, S. Savage-Rumbaugh, and H.S. Terrace for constructive comments on an earlier version of this manuscript. The responsibility for opinions and any inaccuracies is, however, the authors' and not theirs. We also thank J. Rosenblatt, editor of Advances in the Study of Behavior, for permission to use parts of a forthcoming longer paper to be published by Academic Press.

REFERENCES

(1) Bellugi, U., and Klima, E.S. 1976. Two faces of sign: iconic and abstract. In Origins and Evolution of Language and Speech, eds. S.R. Harnad, H.D. Steklis, and J. Lancaster, vol. 240, pp. 514-538. New York: Annals of the New York Academy of Sciences.

(2) Bennett, J. 1976. Linguistic Behavior. Cambridge: Cambridge University Press.

(3) Bennett, J. 1978. Some remarks about concepts. Behav. Brain Sci. 4: 557-560.

(4) Dennett, D.C. 1969. Content and Consciousness. New York: Humanities Press.

(5) Dennett, D.C. 1978. Intentional systems. In Brainstorms: Philosophical Essays on Mind and Psychology. Montgomery, VT: Bradford Books.

(6) Desmond, A.J. 1979. The Ape's Reflexion. New York: Dial Press.

(7) Farrer, D.N. 1969. Picture memory in the chimpanzee. Percept. Motor Skills 25: 305-315.

(8) Fouts, R.S.; Chown, W.; and Goodin, L.T. 1976. Transfer of signed responses in American Sign Language from vocal English stimuli to physical object stimuli by a chimpanzee (Pan). Learn. Motiv. 7: 458-475.

(9) Gardner, B.T., and Gardner, R.A. 1974. Comparing the early utterances of child and chimpanzee. In Minnesota Symposium in Child Psychology, ed. A. Pick, vol. 8, pp. 3-23. Minneapolis, MN.; University of Minnesota Press.

(10) Gardner, B.T., and Gardner, R.A. 1975. Evidence for sentence constituents in the early utterances of child and chimpanzee. J. Exp. Psychol. General 104: 244-267.

(11) Gardner, B.T., and Gardner, R.A. 1979. Two comparative psychologists look at language acquisition. In Children's Language, ed. K.E. Nelson, vol. 2. New York: Gardner Press

(12) Gardner, R.A., and Gardner, B.T. 1969. Teaching sign language to a chimpanzee. Science 165: 664-672.

(13) Gardner, R.A., and Gardner, B.T. 1978. Comparative psychology and language acquisition. In Psychology the State of the Art, eds. K. Salzinger and F.L. Denmark vol. 309, pp. 37-76. New York: Annals of the New York Academy of Sciences.

(14) Grice, H.P. 1967. Logic and conversation. William James Lectures, Harvard University. In Studies in Syntax, eds. P. Cole and J.L. Morgan, vol. 3. New York: Academic Press, 1975.

(15) Hayes, C. 1951. The Ape in Our House. New York: Harper.

(16) Held, R., and Hein, A. 1963. Movement-produced stimulation in the development of visually guided behavior. J. Comp. Physiol. Psychol. 56: 872-876.

(17) Kellogg, W.N., and Kellogg, L.A. 1933. The Ape and the Child: A Study of Environmental Influence Upon Early Behavior. New York: Whittlesey House. (Reprinted 1967, New York: Haffner.)

(18) Lenneberg, E.H. 1975. A neuropsychological comparison between man, chimpanzee and monkey. Neuropsychologia 13: 125.

(19) McDermott, D. 1981. Artificial intelligence meets natural stupidity. In Mind Design, ed. J. Haugeland. Montgomery, VT: Bradford Books.

(20) McNeill, D. 1974. Sentence structure in chimpanzee communication. In The Growth of Competence, eds. K. Connolly and J. Bruner, pp. 75-94. New York: Academic Press.

(21) Menzel, E.W. 1974. A group of young chimpanzees in a one-acre field. In Behavior of Nonhuman Primates, eds. A.M. Schrier, and F. Stollnitz, vol. 4, pp. 83-153. New York: Academic Press.

(22) Miles, H.L.W. 1978. Conversations with apes. The use of sign language by two chimpanzees. Ph.D. dissertation. University of Connecticut.

(23) Nelson, K. 1973. Structure and strategy in learning to talk. Monog. Soc. Res. Child Dev. 38: 1-137.

(24) Patterson, F.G. 1978. The gestures of a gorilla: Language acquisition in another pongid. Brain Lang. 5: 72-97.

(25) Patterson, F.G. 1979. Linguistic capabilities of a lowland gorilla. Ph.D. dissertation, Stanford University. (University Microfilms International Edition, Ann Arbor, MI)

(26) Premack, D. 1976. Intelligence in Ape and Man. Hillsdale, NJ: Lawrence Erlbaum.

(27) Premack, D., and Woodruff, G. 1978. Does the chimpanzee have a theory of mind? Behav. Brain Sci. 4: 515-526.

(28) Ristau, C.A., and Robbins, D. 1979. A threat to man's uniqueness? Language and communication in the chimpanzee. J. Psycholing. Res. 8: 267-300.

(29) Ristau, C.A., and Robbins, D. 1981. Language in the great apes: A critical review. In Advances in the Study of Behavior, eds. J. Rosenblatt, R.A. Hinde, C. Beer, M.-C. Busnel, vol. 12. New York: Academic Press.

(30) Rumbaugh, D.M., ed. 1977. Language Learning by a Chimpanzee: The Lana Project. New York: Academic Press.

(31) Rumbaugh, D.M., and Gill, T.V. 1977. Lana's acquisition of language skills. In Language Learning by a Chimpanzee: The Lana Project, ed. D.M. Rumbaugh, pp. 165-192. New York: Academic Press.

(32) Ryle, G. 1949. The Concept of Mind. New York: Barnes & Noble.

(33) Savage-Rumbaugh, E.S.; Rumbaugh, D.M.; and Boysen, S. 1978. Linguistically mediated tool use and exchange by chimpanzees (Pantroglodytes). Behav. Brain Sci. 4: 539-554.

(34) Savage-Rumbaugh, E.S.; Rumbaugh, D.M.; and Boysen, S. 1980. Do apes use language? Am. Sci. 68: 49-61.

(35) Savage-Rumbaugh, E.S.; Rumbaugh, D.M.; Smith, S.T.; and Lawson, J. 1980. Reference-the linguistic essential. Science 210: 922-925.

(36) Seyfarth, R.M.;. Cheney, D.L.; and Marler, P. 1980. Monkey responses to three different alarm calls: evidence of predator classification and semantic communication. Science 210: 801-803.

(37) Terrace, H.S. 1979. Is problem-solving language? J. Exp. Anal. Behav. 31: 161-175.

(38) Terrace, H.S. 1979. Nim. New York: Knopf.

(39) Terrace, H.S.; Petitto, L.A.; Sanders, R.J.; and Bever, T.G. 1979. Can an ape create a sentence? Science 200: 891-902.

(40) van Lawick-Goodall, J. 1971. In the Shadow of Man. Boston, MA: Houghton-Mifflin.

(41) Woodruff, G., and Premack, D. 1979. Intentional communication in the chimpanzee: the development of deception. Cognition 7: 333-362.

(42) Zaidel, E. 1978. Of apes and hemispheres. Behav. Brain Sci. 4: 607-609.

Group on Neuropsychological Approaches

Standing, left to right: Rudolf Cohen, Steve Hillyard, Bill Hodos, Floyd Bloom, Ted Bullock, Andreas Elepfandt, Gary Lynch.
Seated: Helen Neville, Jerre Levy, Patricia Churchland Smith, Pat Goldman-Rakic.

Animal Mind - Human Mind, ed. D.R. Griffin, pp. 333-353.
Dahlem Konferenzen 1982. Berlin, Heidelberg, New York: Springer-Verlag.

Neuropsychological Approaches
State of the Art Report

H. J. Neville and S. A. Hillyard, Rapporteurs
F. E. Bloom, T. H. Bullock, R. J. Cohen, A. Elepfandt,
P. S. Goldman-Rakic, W. Hodos, J. Levy, G. S. Lynch,
P. Smith Churchland

INTRODUCTION

What are the critical characteristics of structure, function, and organization that an animal's brain must possess in order to carry out activities that are undeniably "mental"? Are there particular anatomical, physiological, or chemical features of the brain that are reliable markers of specific mental functions or conscious experiences? Do different species exercise radically different modes of consciousness and mental states corresponding to species-specific forms of brain organization? These are some of the principal questions that guided the discussions of our group, whose members shared the view that it is possible, at least in principle, to translate mental phenomena into the realm of neural mechanisms, however complex these may turn out to be. To the extent that we can identify the essential neural bases of mental and conscious states, we will then be able to make comparative evaluations of the mental lives of animals using neuropsychological approaches.

Before discussing the various lines of neuropsychological experimentation that are pertinent to these questions, the group attempted to arrive at some consensus as to the definitions of the terms "mind" and "consciousness." How should we characterize the mental phenomena for which we are seeking neural bases, and indeed, how would we know when such are present? After much debate, it was decided that we are not yet in a position to pose formal definitions of "mental" or "conscious" phenomena. While this situation hinders the focusing of research upon specific hypotheses, it is typical of the early stages of any scientific discipline that the phenomena under investigation can only be specified in a rudimentary fashion. As understanding develops, the phenomena will come to be defined more analytically and their definitions embedded in more rigorous theoretical frameworks. However, experimental investigations should not be held in abeyance because of weak theoretical formulations, but should provide new evidence for their refinement.

In this context, we were reminded that the properties of light continued to be studied for many years after Newton and Huygens formulated contradictory theories as to its nature, both broadly derived from the ancient Greeks. The parochial notions of waves and corpuscles were sufficient for guiding considerable experimental progress in these earlier eras of physical optics. Today, the behavior of light is more precisely understood, but its "nature" is difficult to comprehend at the level of intuition or common sense - it exhibits properties of both waves and particles.

While our conceptions of "mind" and "consciousness" may be similarly intuitive at this stage, working definitions or criteria are needed to focus research and to promote new hypotheses. Further refinement of these concepts will go hand in hand with experimental progress. Thus, the group suggested a few rough guidelines or criteria for assessing whether processes may be "conscious" or "mental":

1. The process should occur in the nervous system of a biological organism.

2. Neural events that are triggered reflexively and automatically (e.g., knee jerk) are less likely to engender conscious experience than those which are actively controlled (e.g., suppression of knee jerk).

3. Behaviors that are flexible and adaptively responsive to environmental contingencies are more likely to occur under conscious control than those which are pre-programmed. In particular, studies of the mechanisms underlying the acquisition of complex skills should allow comparisons of conscious regulatory processes in different species.

4. Similarities of brain structure and function between animals and the human adult (the paradigmatic case for our intuitive notions of minds) were taken by some group members as a basis for inferring mental processes. However, Bullock warned that the relative size of the cerebral cortex cannot be assumed to indicate in any simple proportionality the degree of mental life or self-awareness among various mammals. Similarly, Hodos suggested that highly complex brains that are configured with little or no cortex might also support forms of conscious awareness.

5. Since the mental repertoire of the adult human emerges through a well-defined sequence of developmental stages, comparative studies of cognitive and behavioral development could provide further evidence for designating mental processes in animals.

6. Since human language is closely coupled to thought and consciousness, animals which display similar communicative systems seem more likely to harbor human-like mental processes (see Seyfarth et al., this volume).

The thorny issue of the status of "unconscious processing" was also discussed. While a conscious thought or image is a mental process *par excellence*, what is the status of an unconscious thought? Following the lead of Helmholtz and Freud, philosophers and psychologists have come to accord mental status to covert events that cannot be reported, by virtue of their theoretical relations to conscious content or behavior. Indeed, present-day cognitive theorists propose the existence of unconscious events such as "template matching" (in recognition), "retrieval" (in remembering), or "search" (in trying to remember). Thus, it seems likely that many functions of animal minds could be carried out unconsciously and yet be considered full-fledged mental events.

With these intuitive and denotative criteria in mind, the group considered how animal minds might be explored through behavioral, neuroanatomical, and physiological analyses of brain function.

BEHAVIORAL ANALYSES OF BRAIN ORGANIZATION

A remarkable and significant feature of the human brain is its asymmetrical functional organization. Behavioral studies of intact and brain-damaged humans have repeatedly shown that in the adult the two cerebral hemispheres are specialized for different mental abilities including linguistic, cognitive, attentional, and perceptual functions. In most right-handed adults language material is processed more accurately after presentation contralateral to the left hemisphere (i.e., to the right ear, visual field, or hand) and the production and comprehension of language is far more likely to be disrupted by lesions to the left than the right hemisphere. On the other hand, processing of certain types of non-language material, especially those requiring the perception of spatial relationships, has been shown to be more dependent on the right hemisphere (see Levy, this volume).

For those who view speech as a prerequisite for consciousness, the differential functional specializations of the hemispheres becomes a possible neural substrate of consciousness. If lateral specialization can be shown to be a reliable correlate of consciousness, future research should focus on mental events of organisms displaying such specialization. While the evidence for lateral specialization in other primates has only been modest (see Hillyard and Bloom, this volume), a pattern of functional asymmetries remarkably similar to that for human language seems to underlie the production of song in certain birds (38). Moreover, in birds which have lateralized control of singing, the songs are to a large extent learned. These data raise the question of the significance of brain asymmetry for learning and information processing of the type required in human speech and bird song. Are these functional asymmetries closely related or are they merely examples of convergent evolution? Both forms of communication require the processing of acoustic sequences rapidly modulated in time. Behavioral studies of humans have shown that the perception of temporal order of rapidly modulated non-language acoustic sequences also appears to be mediated by the left hemisphere (28) while spatial processing is predominant in the right hemisphere. These results have suggested the hypothesis that cerebral specialization in birds may be similarly characterized in terms of distinct temporal and spatial processing modes. Moreover, it is conceivable that similar patterns of hemisphere specialization exist in other species that process complex spatial and temporal information (e.g., bats).

While most investigations of lateral asymmetries have focused on differential specializations for cognitive abilities, some behavioral investigations of intact and brain-damaged humans have also revealed asymmetries in affective processes. For example, greater emotional expression has been observed after presentation of faces to the left visual field in humans (11), and lesions to the left and right hemispheres have been reported to produce pathological depression and elation, respectively (16). It has also been reported that unilateral hemispheric lesions produce distinct effects on emotionality in rats (14).

Certain interspecies parallels in the emergence of specialization of function during ontogeny are evident. For example, in man, the functional differentiation of the two hemispheres is not fixed until maturation of the brain is largely completed (by several neural indices (33)). In birds, a similar course of lateral differentiation for song is observed (37). Moreover, in man, several types of studies (ERPs, lesion, and behavioral) suggest the importance of the specific type of language experience (as in the deaf and bilinguals) in determining the nature of lateral specialization ((1,35,36) and Neville, Kutas and Schmidt, submitted). In rodents, certain types of early experience may also determine lateral specialization for emotionality (13).

In addition to the lateralized pattern of cerebral specialization in man, different areas within each cerebral cortex subserve aspects of attentional, mnemonic, and perceptual performance. Other primates also display specific intra-hemispheric specializations for similar mental abilities (see Hillyard and Bloom, this volume). Moreover, there are striking ontogenetic parallels between man and other primates in the timetables of development of functional specializations and the associated periods of plasticity. Thus, both in man and rhesus monkeys, cortical specializations associated with several cognitive abilities are not functionally mature until at least puberty (2,3,5,18,27).

These data are consistent with the notion that a protracted postnatal course of neural maturation is associated with increasing cortical specialization, in turn associated with the emergence of higher mental abilities. Moreover, prolonged development provides a greater potential for environmental shaping, such that more complex, later developing neural systems reveal greater functional plasticity during ontogeny (20,21,25). For example, while the effects of cortical lesions are often short-lived during ontogeny in our own and other primate species, the effects of subcortical lesions are often irreversible (19,24,44).

A dominant view of human development, associated with Piaget (42), is the epigenetic unfolding of cognitive abilities according to psychobiological principles. Future research could be directed to detailing the maturational sequences of cognitive abilities and associated neural structures in our own and closely related species. In particular, the neural reorganization and functional recovery that occurs in human infants after neurological insult may be reflected in other primates, providing a further avenue for relating intellectual processes and structures.

In summary, several lines of investigation suggest that increasing complexity of animal mind may be associated with 1) an increasing role of the cortex in mediating behavior and mental events (but see Hodos, this volume); 2) increasing functional specializations between and within the cerebral hemispheres; and 3) prolongation of postnatal neural maturation, and associated sensitive or critical periods when specific functions emerge in association with appropriate stimulation and with the readiness of relevant brain systems.

Behavioral Reactions to Drugs

Comparing the effects of drugs on the behavioral reactions of animals and people can provide another approach for assessing the animal mind. To the extent that a particular class of mental processes shares common or homologous neurophysiological and neurochemical bases in different species, the impact of "psychoactive" drugs on its behavioral manifestations should be similar. The effects of pain-killing drugs illustrate this approach. In general, drugs which alleviate the behavioral and psychological reactions to painful stimuli in humans also produce corresponding reductions in standard behavioral indices of pain in animals (43). While this evidence in itself would not be sufficient to conclude that animals experience pain in the same way people do, it is a first step in that direction. If the drug effects could be equated on a number of behavioral dimensions and on psychophysiological measures as well (e.g., effects on evoked potentials and autonomic responses), the case for a cross-species similarity of pain mechanisms would be strengthened.

Bullock pointed out that some indices of pain would be more convincing than others in establishing cross-species similarities. Thus behavioral dimensions that measure withdrawal would not distinguish between a person in high level conscious suffering and a person or a paramecium or a spinal frog in a low level or unconscious state who may exhibit similar movements. The challenge is to distinguish between simple reaction to noxious stimuli and suffering with high level affect.

Further converging evidence on the nature of pain in animals could be developed from brain lesion experiments. In man, prefrontal lobotomy has been shown to reduce the subjective intensity of pain and alter its quality; that is, patients with chronic pain syndromes report that they still "feel" the pain after lobotomy, but it does not "bother them" as much as before (4). If the frontal lobes of other mammals are similarly important for the complete elaboration of the pain experience, this should be evident through careful measurement of their behavioral reactions to noxious stimuli following the ablation of the homologous cortical areas. Possible alterations in evoked potential amplitudes, particularly in those late components that have been correlated with subjective pain ratings in man, would also be helpful in assessing the comparability of the lesion effects.

COMPARATIVE NEUROANATOMY

Most of the gross quantitative measures of brain size that have been related to mental capabilities across species suffer from serious limitations. The standard allometric ratio of brain weight/body weight is less than satisfactory, because body weight is affected by a variety of selective pressures that are not correlated with brain size (e.g., flying animals are under considerable pressure to keep body weight to a minimum). Thus, the functional significance of brain/body ratios is highly uncertain.

The "Encephalization Quotient" (EQ) is another comparative ratio that is difficult to interpret. The EQ for a particular species is the ratio of its brain size to the brain size of the "average" animal of that body weight as specified by the overall regression line relating brain to body weights across a large number of species (30). Thus, the EQ measure registers proportional variations in brain size rather than absolute variations; in order for a large-bodied creature to achieve as high an EQ as a small-bodied creature, a much larger absolute increment in brain weight above normal regression line is required. Jerison (30) has suggested that the total number of "excess neurons" above the "average" would be a more valid measure of the relative sophistication of species' processing capacities.

Some interesting relationships emerge when the sizes of particular brain structures are plotted against whole brain size in different species. Passingham (40) has compiled these measurements for a number of primate species including man. He finds that the association areas of the cortex increase in size much more steeply as a function of whole brain size than do, say, the cerebellum or the caudate nucleus. This highlights the increasing importance of association cortex in the larger primates. Most significantly, however, the brain structures of Homo sapiens are specified by exactly the same regression lines as fit the other primates. There is no abrupt discontinuity in the relative sizes of these brain structures in man. Our own species shows a direct extrapolation of the basic primate pattern, at least insofar as these crude weight measures are concerned. However, an exaggerated increase in the proportion of cortico-cortical connections has been reported in humans in relation to other primates (31).

The relative size of different brain structures varies widely in different species in a way that frequently can be related to their specialized ecological adaptations. The size of those brain regions which subserve specific sensory, motor and associative functions are magnified to the extent that those functions are exercised by the species in question. This principle appears very general across different functional systems and different levels of brain organization. An interesting implication of this interspecific diversity is that qualitatively distinct mental processes and/or states of consciousness may have evolved in parallel with specialized modes of information processing.

A survey of the major regions of the central nervous system (see Hodos, this volume) indicated that certain structures had achieved particularly striking development in different taxonomic groups and suggested which of these might likely be involved in mental activities and intelligent behavior. The spinal cord has traditionally been ruled out in this regard, but it was pointed out that some of the nociceptive reflexes of the frog spinal cord are quite sophisticated in their responsiveness to stimulus contingencies. The cerebellum also was deemed to be an unlikely participant in mental activity, both because its destruction produces a negligible effect on human cognitive functioning and because its organizational plan and types of cells are highly similar across all vertebrate classes. Some species of rays (e.g., manta) and of bony fishes (e.g., mormyrids) have an extraordinarily large cerebellum; in the latter the cell organization is quite divergent from the usual. Attempts to relate these special cases to electroreception encounter problems not yet solved. Similarly the relatively very well developed cerebellums of sloths, manatees, and whales - as large in proportion as those of deer, panther, and monkey, underline our ignorance of the meaning of this organ and of size.

The cranial nerves are also organized along essentially similar lines in most vertebrates. However, certain brain structures which receive input from cranial nerves are highly developed (in relation to humans) in species which make use of specialized sensory inputs. Examples are the great enlargement of the vagal lobes in fish that have dense arrays of external taste buds, the elaboration of the trigeminal projections in reptiles with infrared sensitivity, the enlarged olfactory lobes in animals that depend on their sense of smell. Conceivably, some of these uniquely evolved sensory systems could engender their own special forms of perceptual experience in the animals.

While inter-species variations in regional brain development can usually be directly related to functional specializations, in some cases this link is not so evident. For example, while both dolphins and bats rely on echolocation for orientation and have highly developed auditory nuclei in their midbrains, the dolphin has a much larger nucleus of the lateral lemniscus than the bat. This makes one wonder whether some acoustic processing functions may be added or exaggerated in the dolphin that would be apparent if their hearing were to be evaluated with more sensitive behavioral tests.

The tectum plays a major role in many vertebrate classes in representing spatial relationships between an animal and its external environment. Variations in tectal size and development across species seem to correlate with the sophistication of spatially-directed behaviors, suggesting that tectal mechanisms might form the basis of an awareness for the spatial properties of stimuli. Since the tectum evidently plays a lesser role in visual processing in species with highly developed geniculo-cortical systems, however, its contribution to sensory awareness might well be species-specific.

Relevant to this question of tectal functions is the phenomenon of "blind sight" that is seen in some human patients following lesions of the visual cortex that produce complete blindness as assessed by conventional tests (41,45). When examined more closely, it has been found that such patients may respond accurately to the location of objects in "blind" portions of their visual fields and may discriminate certain visual texture attributes, while denying any conscious perceptual awareness of the stimuli. This suggests that in species like ours, where vision is highly corticalized, that subcortical structures including the tectum can process visual inputs and organize discriminative responses without producing the conscious sensation of "seeing" the stimulus.

Inter-species variations in limbic structures have naturally been related to the richness of an animal's emotional experiences. On these grounds, one might expect emotional awareness to be extensively distributed throughout the vertebrates, since most orders have a well-defined septal nucleus, amygdala, and hypothalamus. Some limbic structures, however, notably the hippocampus, only become prominent in mammals. Given the purported role of the hippocampus and associated temporal lobe structures in higher memory functions, particularly memory for spatial locations, it is little wonder that this impressive structure has been targeted as a possible center for emergent cognitive processes (39).

In general, cross-species similarities in the localization of cognitive functions to specific brain regions were considered to provide a basis for inferring mental qualities to animals (see Hillyard and Bloom, this volume). Many techniques are currently available for comparative studies of functional localization, including cryogenic lesions, spreading depression, electrical stimulation, and 2-deoxyglucose mapping of regional metabolic activity.

Studies using new, sophisticated neuroanatomical techniques have uncovered more refined aspects of cortical structure that may be relevant for understanding its information processing capabilities (26). The organization of cortical neurons into functional columns has been well-documented in the sensory receiving areas of many mammalian species, and a similar columnar structure has recently been found in the association cortex of monkeys as well (22,23). The concept was proposed that the cortical columns are functional modules that enable an increased number of cortico-cortical fibers and interconnections to be accomodated efficiently (26). The columnar plan could go hand in hand with an increasing need to process more complex types of information from multiple sources. Inter-species variations in other architectural features such as laminar and modular organization could provide a basis for emergent functions (8,9). However, these powerful neuroanatomical techniques cannot be applied to study human brains since they require lesions and/or labelled injections in vivo, and subsequent histological analysis of prepared tissue.

PHYSIOLOGICAL APPROACHES

A number of types of event-related potentials (ERPs) can be recorded in the human EEG in association with cognitive, perceptual, and linguistic processing (10). These ERPs represent the summated electric fields that arise from the temporally and spatially patterned neural activity that accompanies these psychological events. Typically, ERPs are elicited by sensory stimuli and are extracted from the background EEG by computer.

Although the anatomical and physiological bases of these ERPs are largely unknown, some of them are highly reliable correlates of processes that are generally considered to be mental and involving states of conscious awareness when they occur in adult human beings. This suggests a research strategy for investigating the extent to which comparable mental processes might exist in animals and man. If careful behavioral-physiological studies in man have shown an ERP to be a reliable marker of a more or less well-defined mental process, then the demonstration of a comparable ERP configuration in a nonhuman species under appropriate circumstances would support the proposal that similar brain circuitry was being used by both species to accomplish that process. Making the further assumptions, that comparable neurophysiological patterns imply comparable mental events (mind-body identity) and that scalp recorded ERPs reflect essential aspects of this underlying neurophysiology, inferences may be made about the possible equivalence or nonequivalence of mental events in man and animals (see Hillyard and Bloom, this volume).

The ERPs which have been best characterized as to their psychological correlates include the following:

1. The N100 component (a negative wave of maximum amplitude around 100 msec after stimulus presentation) elicited by stimuli belonging to a selectively attended channel of input. N100 is largest when a person must focus attention on one channel in a "noisy" environment (e.g., one voice at a cocktail party (29)).

2. The P300 component, a positive wave which is triggered when an unexpected or surprising stimulus occurs (15). There appear to be two classes of P300 waves. The first is largest over the parietal scalp and is elicited by a relevant stimulus that requires some sort of reaction (overt or covert). For example, a P300 follows the detection of threshold-level signals in a detection task and its amplitude grows in proportion to the confidence of the detection (clarity of the perception). The P300 also accompanies false positive reports of signals that are not present, indicating that this ERP is more closely coupled with perceptual processes than with peripheral sensory input. Numerous studies have shown that the P300 is entirely dissociable from motor response processes (15). The second type of P300 is seen in response to novel or irrelevant stimuli that surprise the subject. This P300 has a more anterior brain distribution and habituates upon repetition of the novel event.

3. The N400 wave is elicited under conditions where a verbal or meaningful stimulus violates a semantic context (32). For

example, an N400 follows the semantically inappropriate word sentences such as "For breakfast this morning I ate eggs cigars." An N400 wave also follows pictures which are out of context, such as a picture of an automobile in a verbal context which describes a garden.

4. Finally, the CNV or expectancy wave is a slow negative ERP which occurs preceding anticipated stimuli (15). This ERP has also been called the "intention" wave, since it occurs during the period when a person is preparing to make a motor act to an expected stimulus.

All of these ERPs occur to stimuli that are processed with conscious awareness, at least in the usual non-rigorous sense of the term. Persons who have just produced N100 or P300 waves to stimuli are able to give a full verbal description of what they experienced, as well as make appropriate discriminative responses. These waves are diminished with drowsiness or distraction. While each class of ERP seems to be associated with a qualitatively different perceptual or cognitive process, the attribute of awareness seems present in each case.

To proceed with the analysis of mental events in animals using ERPs as markers, two types of experiments need to be pursued: The first would be continuing studies in humans to define more precisely the psychological correlates of these ERPs and others that are also being studied. Griffin asked, for instance, whether there would be a P300 wave when a relevant signal was detected and responded to automatically, without conscious awareness. This question is amenable to investigation by presenting subjects with a very demanding task that would divert their attention from a detection task that would normally elicit a P300.

Second, experiments are needed which make explicit attempts to identify animal homologues of the aforementioned human

ERPs (17). While numerous studies have examined electrophysiological responses in the brains of alert, behaving animals, few have presented stimuli in tasks that are equivalent to those in which the human ERPs are observed. Only recently have animal studies been designed with the aim of eliciting P300-like waves, and promising candidates for P300 homologues have been observed in both cats (46) and squirrel monkeys (Neville and Foote, in preparation).

Another promising line of research is on the ontogenetic development of the long latency ERPs, which show clear age-dependent changes in morphology and scalp distribution in man (12). It seems reasonable to expect that homologous ERPs in animals would show similar developmental changes.

Cortical information processing may also be described in terms of the synaptic organization and neurotransmitter actions within neuronal ensembles. Several types of neurotransmitter actions have been characterized, including the familiar excitation, inhibition, and electrical interactions. More recently, however, certain neurotransmitters have been shown to have "enable" (monoamines) and "disenable" (peptides) functions. An enabling synapse, when active, potentiates <u>both</u> the excitatory and inhibitory actions of other inputs to the neuron, while a disenabling synapse suppresses other inputs. The monoaminergic fibers (e.g., from the locus coeruleus) have widespread cortical projections and intercept cortical columns in such a way that they could determine which columns would be active at a particular time (see Hillyard and Bloom, this volume).

Bloom suggested that the enable-type of synaptic action may be particularly involved with higher levels of information processing (and mentation?), in that this would allow for complex activity patterns to be sustained for appropriately long durations. This corresponds with proposals by Libet (34) that conscious sensations are dependent upon appropriately prolonged neural processing (over several hundred msec). Of interest are recent studies showing considerable differences between rodents and primates in the development (5,27), regional distribution (6), and functional significance (7) of monoamine systems in the cerebral cortex. Hence, further cross-species comparisons of the extent of "enable"-type neural processing may prove instructive in relation to the sophistication of information processing.

It was further noted that the action of the "enable"-type synapse also lasts for several hundreds of milliseconds, comparable to durations of human ERPs. This lends further credence to the hypothesis that the long-latency ERPs in man are manifestations of neural events closely tied to conscious processing.

REFERENCES

(1) Albert, M.L., and Obler, L.K. 1978. The Bilingual Brain, Neurophysiological and Neurolinguistic Aspects of Bilingualism. New York: Academic Press.

(2) Alexander, G.E., and Goldman, P.S. 1978. Functional development of the dorsolateral prefrontal cortex: An analysis utilizing reversible cryogenic depression. Brain Res. 143: 233-250.

(3) Alexander, G.E.; Witt, E.; and Goldman-Rakic, P.S. 1980. Neuronal activity in the prefrontal cortex, caudate nucleus and mediodorsal thalamic nucleus during delayed response performance of immature and adult rhesus monkeys. Neurosci. Abstr. 6: 86.

(4) Barber, T.X. 1959. Toward a theory of pain: Relief of chronic pain by prefrontal leucotomy, opiates, placebos and hypnosis. Psychol. Bull. 56: 430-460.

(5) Brown, R.M., and Goldman, P.S. 1977. Catecholamines in neocortex of rhesus monkeys: Regional distribution and ontogenetic development. Brain Res. 124: 576-580.

(6) Brown, R.M.; Crane, A.; and Goldman, P.S. 1979. Regional distribution of monoamines in the cerebral cortex and subcortical structures of the rhesus monkey: Concentrations and in vivo synthesis rates. Brain Res. 168: 133-150.

(7) Brozoski, T.H.; Brown, R.M.; Ptak, J.; and Goldman, P.S. 1979. Dopamine in prefrontal cortex of rhesus monkeys: Evidence for a role in cognitive function. In Catecholamines: Basic and Clinical Frontiers, eds. E. Usdin, I.J. Kopin, and J. Barchas, pp. 1681-1683. New York: Pergamon Press.

(8) Bugbee, N.M., and Goldman-Rakic, P.S. 1980. Compartmentalization of prefrontal projections: Comparisons of cortical columns and striatal islands in old and new world monkeys. Neurosci. Abstr. 6: 822.

(9) Bugbee, N.M., and Goldman-Rakic, P.S. 1981. Modular organization in cortex and striatum: Comparison of prefrontal columns and striatal compartments in macca and saimiri monkeys. Brain Res., in press.

(10) Callaway, E.; Tueting, P.; and Koslow, S.H. 1978. Event-Related Brain Potentials in Man. New York: Academic Press.

(11) Campbell, R. 1978. Asymmetries in interpreting and expressing a posed facial expression. Cortex 14: 327-342.

(12) Courchesne, E. 1977. From infancy to adulthood: The neurophysiological correlates of cognition. In Cognitive Components in Cerebral Event-Related Potentials and Selective Attention. Progress in Clinical Neurophysiology, ed. J.E. Desmedt, vol. 6, pp. 224-242. Basel: Karger.

(13) Denenberg, V.H. 1981. Hemispheric laterality in animals and the effects of early experience. Behav. Brain Sci., in press.

(14) Denenberg, V.H.; Garbanati, J.; Sherman, G.; Yutzey, D.A.; and Kaplan, R. 1978. Infantile stimulation induces brain lateralization in rats. Science 201: 1150-1152.

(15) Donchin, E.; Ritter, W.; and McCallum, W.C. 1978. Cognitive psychophysiology: The endogenous components of the ERP. In Event-Related Brain Potentials in Man, eds. E. Callaway, P. Tueting, and S.H. Koslow. New York: Academic Press.

(16) Gainotti, G. 1972. Emotional behavior and hemispheric side of lesion. Cortex 8: 41-55.

(17) Galambos, R., and Hillyard S. 1981. Electrophysiological approaches to human cognitive processing. NRP Bull., in press.

(18) Goldman, P.S. 1971. Functional development of the prefrontal cortex in early life and the problem of neuronal plasticity. Exp. Neurol. 32: 366-387.

(19) Goldman, P.S. 1974. An alternative to developmental plasticity: Heterology of CNS structures in infants and adults. In Plasticity and Recovery of Function in the Central Nervous System, eds. D.G. Stein, J.J. Rosen, and N. Butters. New York: Academic Press.

(20) Goldman, P.S., and Galkin, T.W. 1978. Prenatal removal of frontal association cortex in the rhesus monkey: Anatomical and functional consequences in postnatal life. Brain Res. 52: 451-485.

(21) Goldman, P.S., and Lewis, M.E. 1978. Developmental biology of brain damage and experience. In Neuronal Plasticity, ed. C. Cotman, pp. 291-310. New York: Raven Press.

(22) Goldman, P.S., and Nauta, W.J.H. 1977. An intricately patterned prefronto-caudate projection in the rhesus monkey. J. Comp. Neurol. 171: 369-386.

(23) Goldman, P.S., and Nauta, W.J.H. 1977. Columnar distribution of cortico-cortical fibers in the frontal association, motor and limbic cortex of the developing rhesus monkey. Brain Res. 122: 393-413.

(24) Goldman, P.S., and Rosvold, H.E. 1972. The effects of selective caudate lesions in infant and juvenile rhesus monkeys. Brain Res. 43: 53-66.

(25) Goldman-Rakic, P.S. 1980. Morphological consequences of prenatal injury to the brain. In Adaptive Capabilities of the Nervous System, Progress in Brain Research, eds. P.S. McConnell, G.J. Boer, H.J. Romijn, N.E. van de Poll, and M.A. Corner, vol. 53, pp. 3-19. Amsterdam: Elsevier.

(26) Goldman-Rakic, P.S. 1981. Development and plasticity of primate frontal association cortex. In The Organization of the Cerebral Cortex, ed. F.O. Schmitt. MIT Press.

(27) Goldman-Rakic, P.S., and Brown, R.M. 1981. Regional changes of monoamines in cerebral cortex and subcortical structures of aging rhesus monkeys. Neurosci. 6: 177-187.

(28) Halperin, Y.; Nachshon, I.; and Carmon, A. 1973. Shift of ear superiority in dichotic listening to temporally patterned nonverbal stimuli. J. Acoust. Soc. Am. 53: 46-50.

(29) Hillyard, S.A.; Picton, T.W.; and Regan, D. 1978. Sensation, perception and attention: Analysis using ERPs. In Event-Related Brain Potentials in Man, eds. C. Callaway, P. Tueting, and S.H. Koslow. New York: Academic Press.

(30) Jerison, H.J. 1973. Evolution of the Brain and Intelligence. New York: Academic Press.

(31) Katznelson, R.D. 1981. Normal modes of the brain: Neuroanatomical basis and a physiological theoretical model. In Electric Fields of the Brain, ed. P.L. Nunez. Oxford University Press.

(32) Kutas, M., and Hillyard, S.A. 1980. Reading senseless sentences: Brain potentials reflect semantic incongruity. Science 207: 203-205.

(33) Lenneberg, E. 1967. Biological Foundations of Language. New York: John Wiley and Sons.

(34) Libet, B. 1973. Electrical stimulation of cortex in human subjects and conscious sensory aspects. In Somatosensory System, ed. A. Iggo, vol. II. Berlin: Springer-Verlag.

(35) Neville, H.J. 1977. Electrographic and behavioral cerebral specialization in normal and congenitally deaf children: A preliminary report. In Language Development and Neurological Theory, ed. S. Segalowitz. New York: Academic Press.

(36) Neville, H.J., and Bellugi, U. 1978. Patterns of cerebral specialization in deaf adults. In Perspectives in Psycholinguistics and Neurolinguistics, ed. P. Siple. New York: Academic Press.

(37) Nottebohm, F. 1970. Ontogeny of bird song. Science 167: 950-956.

(38) Nottebohm, F. 1977. Asymmetries in neural control of vocalization in the canary. In Lateralization in the Nervous System, eds. S. Harnad, R.W. Doty, L. Goldstein, J. Jaynes, and G. Krauthamer. New York: Academic Press.

(39) O'Keefe, J., and Nadel, L. 1978. The Hippocampus as a Cognitive Map. Oxford University Press.

(40) Passingham, S. 1977. Brain size and intelligence in man. Brain, Behav. Evol. 16: 253-270.

(41) Perenin, M.T., and Jeannerod, M. 1975. Residual vision in cortically blind hemifields. Neuropsychologia 13: 1-7.

(42) Piaget, J. 1971. Biology of Knowledge. Chicago: University of Chicago Press.

(43) Terenius, L. 1978. Endogenous peptides and analgesia. Ann. Rev. Pharmacol. Toxicol. 18: 189-204.

(44) Tucker, T.J., and Kling, A. 1967. Differential effects of early and late lesions of frontal granular cortex in the monkey. Brain Res. 5: 377-389.

(45) Weiskrantz, L.; Cowey, A.; and Passingham, C. 1977. Spatial responses to brief stimuli by monkeys with striate cortex ablations. Brain 100: 655-670.

(46) Wilder, M.B.; Farley, G.R.; and Starr, A. 1981. Endogenous late positive component of the evoked potential in cats corresponding to P300 in humans. Science 211: 605-606.

Group on
Evolutionary Ecology of Thinking

Standing, left to right: John Staddon, Rudolf Drent,
Dietrich Dörner, Franz Reither, Christian Welker, Geoff Parker,
Gordon Orians, John Krebs.
Seated, Jürg Lamprecht, Bill Mason, Marian Dawkins,
Norbert Bischof, Hans Kummer.

Animal Mind - Human Mind, ed. D.R. Griffin, pp. 355-373.
Dahlem Konferenzen 1982. Berlin, Heidelberg, New York: Springer-Verlag.

Evolutionary Ecology of Thinking
State of the Art Report

M. Dawkins, Rapporteur
N. Bischof, D. Dörner, R. H. Drent, J. R. Krebs, H. Kummer,
J. Lamprecht, H. S. Markl, W. A. Mason, G. H. Orians,
G. A. Parker, F. Reither, J. E. R. Staddon, C. Welker

INTRODUCTION

Our group took an initial decision not to discuss the subjective experiences of animals on the grounds that we could see no way of studying them. At first sight, this may seem as though we were avoiding the central issue of the conference altogether, but we took our decision in the belief that by trying to discover the processes, cognitive or otherwise, that lie behind animal behavior, we stood the best chance of being able to generate testable hypotheses. We were thus quite happy to talk about the kinds of cognitive representations that animals might have, but only as long as they could be operationally defined and distinguished by some sort of experiment, however far-fetched. In view of the title given to our group, we took it as our brief to look at the real tasks which animals perform in nature to see if they tell us anything about the rules or algorithms which animals actually use.

We decided that we would make the most progress in this if we concentrated our discussion on specific topics, examples of behavior which seemed to suggest a quite high level of cognitive ability on the part of the animals concerned. The four areas we chose initially were foraging behavior, social relationships, exploration, and the various ways in which animals could be said to evaluate their environments. Each of these generated a number of ideas for experiments and in turn suggested another area we felt we should examine. The internal representations of different animals might well be different and so we felt that it was important to spend some time discussing the evolutionary pressures and constraints that operate on different species. Finally we looked at the problem of studying animal feelings.

ARE THERE EXAMPLES OF ANIMAL BEHAVIOR WHICH SUGGEST COMPLEX COGNITIVE ABILITIES?

Searching for Food

Animals searching their environment for food are confronted with a number of environmental "problems" which seem to require quite high levels of cognitive ability. For example, most animals have to decide where to search. Some species have to avoid going back over the same ground twice, others, on the contrary, to go back to a profitable area, and yet others to go back with a time delay when their food has had time to regenerate. Some concepts of space, time, or combinations of the two would thus seem, on the face of it, to be essential. Is this really so or can the animals get by with simpler mechanisms?

The evidence from maze-learning in rats suggests that at least in this case the animals really do develop a spatial representation of their environment. Rats placed in a maze with eight arms radiating from a central arena can learn to visit each arm just once to find the single piece of food hidden at the end of each arm ((3,13,14,15), Dale and Innis, in preparation). The rats somehow learn not to go down the same arm twice. The most obvious explanation, that they use smell "labels" to tell them which arms they have already visited, can be ruled out by rotating the arms. The possibility that they are systematically searching each arm (for example, going round the maze clockwise) can also be ruled out. The rats appear to have a genuine spatial representation of the maze and then use visual or other cues to tell them which part of the maze corresponds to a particular place on their cognitive "map." Thus, although visual cues are important (as the rats become confused if they are altered too much), the rats are not simply using specific visual landmarks as beacons. This is shown by the behavior of rats which have been blinded after learning a maze using visual cues. Such rats relearn the maze very quickly, clearly making use of their knowledge of the layout of the maze and attaching a new set of cues to their preexisting "map" ((3), and Dale and Innis, in preparation).

To use this spatial map to avoid entering an arm of a radial maze twice, rats have to have a response rule. The particular one they seem to use is to put a temporal "tag" on any arm which they visit and then to choose next the arm with the weakest "tag." In fact, the "least-recent choice" rule turns out to be a good approximation of their observed behavior.

Somewhat similar to the maze-learning of rats is the ability of marsh tits (Parus palustris) to remember where they have hidden each of several hundred food items such as seeds in their territories (2,23). In aviary experiments, marsh tits return to the exact sites where they have earlier stored a seed, and the chance of visiting these sites is much greater than that for very similar nearby sites which were not used for storage (Shettleworth and Krebs, unpublished results). As with Olton's radial maze experiment, suitable controls have eliminated the possibility that the birds relocate their seeds either by direct cues (such as scent) or by a systematic response strategy. One such experiment involved allowing marsh tits to store food with one eye covered and then allowing them to recover seeds at a later stage with either the same or the contralateral eye. With the contralateral eye, the birds were much less likely to visit their storage sites, suggesting both that there is little interocular transfer of memory and that memory is important in the recovery of stored seeds (23).

As we had discussed in connection with the rats, we felt it was important to make a distinction between the animals' representation of its world (its map) and what it does with its map (its response rule). In marsh tits, the response rule might be to return to the last items first, since the most recently stored seeds have the highest probability of having still survived the ravages of decay or depredations by animals other than the one that stored them. On the other hand, if the birds store the most valuable (largest or in best condition) seeds first, they may tend to recover these before the later seeds and therefore show a primacy effect. The important point here is that functional arguments can be used to make predictions about how memory may be organized in both time and space.

This theme, the importance to animals of having information both about time and about space was taken up by Drent in a description of the way parent starlings collect food for their young (this volume). The birds appear to know that certain prey items can be caught only at certain times of day. Shorebirds can anticipate when low tide will occur and they arrive at the wrong time if the wind makes the tide late. Lindauer pointed out that bees, too, can learn to go to different places at different times of the day. Despite their interdependence, however, space and time are separate concepts for animals and can be disentangled experimentally.

Throughout our discussion, we referred to spatial and temporal "maps" and we would like to make it clear that we were not using these terms to mean just any internal representation. An animal might simply respond to landmarks specific to a particular place, but we would not say this animal had a map unless, in addition, the animal could be shown to have some knowledge of the relationships between elements. The animal would have to show evidence, for example, of being able to work out short-cuts in a maze where it previously only followed the longer route (cf. (26)). Lindauer and Gould provided evidence that bees can take short cuts when barriers are removed. We later agreed, at the suggestion of Bischof, that saying that an animal has a map implies two things. There must be some kind of address (in memory) for each element. This would be a location if the map is a spatial one or an individual if one is talking about a social map. There must also be some means whereby information about relationships between elements can be extracted from the set of addresses. We felt that in practice, foraging behavior could potentially provide very valuable insight into the kinds of maps that animals have. Maze learning and food hoarding had already been discussed. Another experiment for the future might be to train a hummingbird to a complex regime of coming to particular places after varying time delays. Since hummingbirds naturally feed on nectar which becomes depleted as they feed on it and then needs time to replenish, they might be able to build up a very complex picture. And what that picture might be could be experimentally determined.

Social Relationships

Kummer (this volume) suggested that the study of social relationships might be another fruitful area for studying the internal representation which animals have of the external world. We might ask some apparently simple questions about what an animal's own concept of group membership consists of. An animal might, for example, "belong" to the group simply because it responds differentially to individuals on a particular site or to animals all having the same smell or a common behavior trait, or it may have a more complex representation of social structure including its own place in it. Young infant hamadryas baboons, incidentally, do not seem to know which band they belong to and move between the sleeping sites of different bands. The juveniles who occasionally retrieve them do seem to have such knowledge (1). Do baboons know which route the other clans of their band will take on any particular day? The fact that they all end up at the same water hole (having been apart for much of the day) suggests some communication of the day's route at departure (see Kummer). The telling anecdote or natural experiment was discussed as an important piece of evidence. After all, what ethologist has not heard of Tinbergen's sticklebacks responding to the red mail van? At the very least, many anecdotes suggest some further experiments.

The difficulties, but at the same time, the possibilities of disentangling exactly what an animal knows about other animals was discussed in relation to crab-eating monkeys (Macaca fascicularis) by Welker. Here, at least in captive animals, the male and female linear rank order is stable over several years. The rank position of the daughters depends only on the rank of their mothers, independent of age and size; the rank of sons depends both on age and their mother's status.

Experiments which involved separating group members and then reintroducing them have shown that individuals know a great deal about their positions in relation to other individuals (29). Females (either with or without their children) quickly regain their former rank after a separation from their group. But when the three lowest-ranking clans were separated from the main group and the others were then reintroduced in gradually increasing rank order (lowest status first), the rank order was upset. Previously, low ranking individuals were found to assume high rank, except in one instance where three very high-ranking females immediately won back their social positions on reintroduction. This exception was particularly revealing because, in the course of the experiment, many of the other females which had previously lost their rank began attacking the now high-ranking individuals. It was as though they had been waiting for a chance to assume their old positions. This suggests that the monkey's perception of their social rank is more complex than simply which other individuals threaten them. They know both their old social positions and their new ones.

We next discussed the various ways in which social hierarchies can arise. The simplest case is where individuals differ in, say, size, and simply rank themselves according to this variable. This can lead to a hierarchy without any individual recognition, merely the use of some cue which can be "agreed upon" by all members of the population. Game theory models can be used to explain why the cue that is used is sometimes "commonsense" (e.g., largest animal wins) and sometimes completely arbitrary (e.g., the most darkly colored animal wins) (Hammerstein and Parker, in preparation). A second, and slightly more subtle way is the "confidence" effect - animals start out equal, but a chance encounter means that one of them wins. His confidence is thus increased, making him more likely to win next time and so on. Again, a hierarchy can result with no true individual recognition. There does seem to be evidence, however, that juncos (Junco hyemalis) do form hierarchies based on true recognition of other individuals (30). At the very least, the juncos appear to divide other birds into "those above me" and "those below me."

One stage more complex are cases where animals appear to know not just the relationships of other animals to <u>them</u>, but the relation of these other animals to each other. Mason described how Rhesus monkeys are able to use each other as "social tools," making use of their knowledge of relationships to do so. For example, if low-ranking monkey C wants to drink, but higher ranking B is keeping him away from water, C may threaten B while making appeasing gestures to an even higher ranking monkey A. C thus uses A's power over B to get B away from the water. Seyfarth provided evidence that vervet monkeys (Cercopithecus aethiops) seem to know which baby belongs to which mother. Mothers can discriminate the calls of their own babies from those of other infants when the calls are played through a loudspeaker. And if they hear the call of a baby which is not their own, they look toward the real mother.

We discussed various ways in which animals' social knowledge could be investigated. One suggestion was a game called Status in which a chimpanzee would have to sort lexigrams representing other chimpanzees into rank order. It might be possible through this, to learn how the chimpanzee saw itself in relation to others and, more interestingly, how much it knew about the relationships between other chimpanzees. If it could deduce the rank order of two animals that it had never seen together from its knowledge of their separate interactions with other individuals, this would suggest that it had a social "map" in the sense we had previously discussed.

Following these examples of actual behavior, Dörner described a theoretical model of problem solving. He defined a problem as a situation in which an organism has no ready-made learned or innate sequences of actions at its disposal to reach its goal. It must construct a new unit of behavior. To do this, it must have construction rules. The most primitive form of such a construction rule is random permutation of existing behavioral units. But the overt and unrestricted trial and error behavior generated by such a rule is dangerous, very time-consuming, and not very successful. Therefore, there will be an advantage to having this permutative trial and error internalized and restricted by the use of restriction rules. The use of restriction rules is one of the most important features of human thinking. An example of a restriction rule is the use of intermediate goals. If a child has the idea of using its mother as a "tool" to get food, the mother is an intermediate goal for the last goal (to get food). Now the problem is restricted. It is no longer necessary to try everything; it is only necessary to try those behavioral units which might bring the child closer to the mother.

We discussed whether observations of what animals actually do could be used to deduce what their restriction rules might be. Might searching strategies tell us something, for instance? Simply finding that an animal behaves in a certain non-random way did not, however, seem to be enough to say whether it was using a restriction rule, for the animal might simply have learned that some actions were more profitable than others or have an innate tendency to pursue a particular kind of search path. But if these possibilities can be eliminated, non-random searching might be an indicator of the use of internal restriction rules.

Gathering Information

To generate any sort of new rules or algorithms, animals need to collect information from their environment. They must therefore explore and sample. Sometimes, they seem to be searching for something specific, like food. At other times, they seem to be collecting information for no immediate purpose. Lindauer described how bees appear to spend up to 40% of their time simply patrolling around the hive to learn what is needed. Then, if the colony is short of workers for some purpose such as nursing, the workers are able to change roles to fit in with the needs of the colony. Even more dramatic is the behavior at swarming. Scout bees go out looking for a suitable site for a new nest several days before swarming actually occurs and then inform the rest of the colony about the suitability of the sites they have found.

We paid particular attention to cases where animals appear to search most when they have "time on their hands." Drent described how starlings with small broods spend a lot of time feeding in areas of low yield, apparently exploring. When they have to work harder, however (as when an experimenter has fiendishly added extra babies to their own brood), they concentrate on the high yield (convenience food) areas. Wehner's observations on desert ants also suggest that they explore more when the living is easy.

The implication behind these examples is that although gathering information (reducing uncertainty) about the environment is valuable, there is also some kind of "cost" to it, even if only in time. For instance, a bird may have discovered a profitable area in which to forage but it can only discover a better one if some of the time which it might otherwise give to feeding is in fact allocated to having a look elsewhere. One study in which the allocation of foraging time to sampling and exploiting has been quantitatively analyzed involved great tits (Parus major) working for food in an operant situation (11). The birds faced a choice of two sites, offering different, but unknown (to the bird) payoffs. The rewards were pieces of food provided on a variable ratio schedule in which a certain number of responses yields, on average, one reward. Birds in this situation start off by visiting both sites roughly in alternation, but they eventually learn to work almost exclusively in the more profitable (more rewards per response) of the two. The policy which maximizes total rewards per experiment involves calculating the benefits of sampling both sites to acquire information and exploiting the better site once it has been identified. The birds achieve a payoff very similar to that predicted by this optimality argument. However, they presumably do this by means of a relatively simple learning rule rather than by carrying out the complex computations that the experimenter has to go through to find the optimal solution. We discussed various examples of such rules including Herrnstein's matching rule (10), a win-stay-lose-shift rule, and the relative sum payoff rule (9).

Evaluation of the Environment

Having already touched on the idea that animals weigh up various costs and benefits, and we then looked in greater detail at the various ways in which they could be said to evaluate their environments and make decisions. One of the most basic decisions an animal is faced with is that of what is and what is not food. Animals with catholic diets, such as rats, are known to have well-developed mechanisms for learning which substances are toxic (7). Rozin pointed out that they also have to learn that there are many other substances like sand which, although not toxic, are not food either (22). Then, even if an animal knows what food is, it may still have to learn to discriminate it from the background. In this connection we discussed the concept of the Search Image (25,28), the idea that animals may have to learn to see very cryptic food. If animals could see cryptic prey when they had "got their eye in" and learned to break the camouflage, would there be any way of showing this (4)? At first glance, such a subjective phenomenon would seem to be quite beyond our grasp, but Signal Detection Theory (8) was suggested by Staddon to be a way of giving it some kind of predictive power. (We did not do justice to Staddon's suggestion in our discussion. It is explained more fully by Staddon, in preparation.)

Animals make decisions in a much wider sense, too. They decide whether to feed or drink, flee from a predator or stay on their eggs, and so on. Their decisions do not appear to depend on just one particular set of motivational variables (such as those that change when the animal goes without food), but on some kind of assessment of many different kinds of internal and external cues. Thus the animal may have to make some kind of calibration of the relative importance of quite disparate requirements, balancing its need for water against the danger from a predator and so on. Although this sounds like an impossible task, the fact that animals do tend to do just one thing at a time (even though they may be stimulated to do several things) suggests that some such evaluation must be going on. Indeed, animals often appear to be able to assess the risks and benefits of their various possible courses of action to a very high degree. Starlings appear to monitor their own energy expenditure and change their behavior during the course of a day as a result (Drent, this volume).

Reither pointed to a parallel between decision making in animals and in humans. Humans often appear to behave "irrationally" according to what they would be expected to do on classical decision theory. But if one assumes that humans put into their decisions some estimate of their own competence at achieving a particular goal and that they try to avoid uncertainty, their behavior becomes much more understandable. Uncertainty over outcome and loss of control are to be avoided. Taking this into account means that seemingly irrational decisions take on a rationality of their own (18,19). We discussed the possibility that animals might also assess their own competence. A female great tit which lays six eggs tends to be better at rearing six young than nine young (shown by experiments in which eggs have been added (17). A female which lays nine eggs to start with, on the other hand, copes with nine young perfectly adequately. Each female thus seems to lay in accordance with her own competence at rearing young.

At this point, the skeptical reader may wonder quite how all this relates to animal mind. Perhaps we should explain. We had been exploring the general area of decision-making in animals. We should possibly have been rather more explicit in saying that we had been considering two sorts of decision. First, there are short-term decisions that an individual makes from moment to moment. These we called tactical decisions (which direction to fly next and so on). Second, there are long-term, strategic decisions (how many eggs to be laid in a lifetime and so on). Of course, any considerations of what is going on in an animal's mind apply only to the short-term decisions (we are not suggesting that animals necessarily see their lives stretching ahead). So the "rules of thumb" that they actually use are what gives us the best idea of their internal representations of the world. But although these are perhaps the crucial issues for this conference, the long-term strategic decisions are not irrelevant. We were in general agreement that the particular rule of thumb that an animal adopts is itself the product of long-term evolutionary considerations. Natural selection affects the kinds of concepts an animal will have. So animals may not think about their fitness, but fitness affects what they think about.

We would like to emphasize, however, that we were not suggesting that easy allusions to "fitness" are the answer to everything. Just because a particular cognitive ability might be thought to be advantageous, there is no guarantee that it will evolve. To make this point clear, we next considered the evolutionary constraints that might operate on different animals. Cognitive abilities turned out to be mixed blessings.

EVOLUTIONARY CONSTRAINTS AND OPPORTUNITIES

We generally avoided using the word "intelligence" but discussed instead possible ways in which this rather general term could be subdivided. Orians proposed a scheme which considered the general phenomenon from two different perspectives, the type of discrimination on the one hand and the level of skill or ability on the other. For example, types of discrimination might include discrimination of conspecifics, either as classes of individuals or as unique individuals, discrimination of non-conspecifics, such as prey, predators, and competitors, and discrimination of environmental features, such as maps in space and time. These categories could evolve more or less independently, so that an animal might have considerable discriminatory abilities with respect to food types but be quite incapable of distinguishing conspecific individuals from one another, and vice versa.

Selecting the most useful categories of abilities (skills) is a bit more difficult because we are at present unable to relate any of the observed behavioral skills with any known neural machinery. Therefore, any such categories must at present be provisional and based upon those behavioral features we are able to measure. Among the candidate categories might be discrimination abilities, storage capabilities, and computational abilities, the latter including the ability to combine previously unassociated computational programs. Such a classification would emphasize the multifaceted nature of "intelligence," while at the same time it would provide a means of progressing, by the use of flexible categories, toward a better understanding of the overall phenomenon.

While the benefits of increased cognitive ability appear fairly obvious (new associations can be formed, new foods exploited, and so on), the costs may be less obvious. However, there may well be considerable construction and maintenance costs. Evolving the relevant circuitry may be expensive (the head may have to get bigger, for example), and learning may be dangerous in itself. An animal engaged in a learning task, or even just exploring, exposes itself to risks of predation and its attention is diverted away from other activities. The costs of the "error" part of trial and error may be heavy, death, for example, through not having yet learned what a predator is. Indeed if such costs are too great, it may be more efficient to be "hard-wired." Even having too much insight and knowledge may be disadvantageous. An animal that knows it is going to deceive another may betray this fact to it and be acted against accordingly. Self-deception (or lack of self-knowledge) may be the best disguise (27).

Phylogenetic constraints may also be important. If an animal does not have receptors that give it, say, stereoscopic information, this may limit what concepts it can form. Each species brings to a situation certain restricted computational abilities which limit the range of solutions it can reach. Some species of birds are extremely good at solving the "string pulling problem" (obtaining food suspended from their perch by a string). These tend to be species which naturally use their feet in feeding and all individuals tend to converge on the same solution. Finches, by contrast, do not use their feet in feeding, have great difficulty with the problem, but eventually solve it - each individual in its own idiosyncratic way (24).

If anyone is tempted to think of natural selection always leading to a perfect solution, they should remember the hedge sparrow, assiduously feeding the young cuckoo which has thrown out the foster parent's own eggs. We discussed this classic example of an animal making a "mistake." It was pointed out that from the cuckoo's point of view, it is not a mistake, and the cuckoo and its host may be engaged in a kind of arms race (6). There is evidence, indeed, to suggest that such an evolutionary arms race has taken place in the past, in that some bird species totally reject the eggs of brood parasites if these are artificially placed in their nests (21). The present-day hosts may be the only species that have not yet outwitted the cuckoo.

The Opportunities

We wondered if we could make any generalizations about the kinds of ecological conditions which are most likely to lead to intelligent behavior in any of its aspects. What effects might group living have? Or hunting active prey? Perhaps in species which have few offspring and prolonged parental care, there might be more opportunities for learning than in species which have many offspring and little parental care. (It might be thought that these two kinds of species corresponded to the K-selected and r-selected categories of MacArthur and Wilson (12); however, this latter distinction was disliked by a number of group members, and we did not feel that it helped us to frame the questions we were interested in.) Mason pointed out, however, that play and exploratory behavior do seem to be characteristic of animals with long periods of immaturity. We felt that the developmental aspects of cognitive abilities were extremely important and that our discussion could have benefited if we had put more emphasis on this topic.

In general, we found it very difficult to survey the animal kingdom and reach any very definite conclusions about the situations most likely to lead to the evolution of highly developed cognitive abilities. Variability of the environment was one possible factor we touched upon. So was complex social structure, although the existence of the termites indicates that eusociality is not necessarily related to learning ability.

Reciprocal altruism (27) was then proposed as one of the most likely settings for the evolution of cognitive abilities. Parker emphasized that reciprocal altruism differs from mutualism in that there is a time delay - the altruistic act is repaid some time later. This means that the altruist must keep track of who is and who is not paying him back, otherwise cheats (who receive benefit but never give it to others) do better than anyone. This in turn leads to a kind of intellectual arms race, on the one side to detect cheating and on the other to cheat undetected. Several difficulties with reciprocal altruism were discussed, one being that the only reasonably documented case is the one of male baboons which are reported to engage in reciprocating pacts to help each other gain access to females (16). Even here, it is not clear whether this is an example of real reciprocal altruism or kin selection. Indeed the distinction between the two appeared to break down the more we discussed it. The altruist must, for the theory to work, direct his altruism selectively to other altruists which will tend, at least initially, to be kin.

Kummer then challenged the assumption which we (and other groups) had been making about reciprocal altruism necessarily involving massive feats of memory and bookkeeping on other individuals. He suggested a much simpler and highly plausible alternative. Animals might come to "value" individuals to whom they had been altruistic and follow them around, thus increasing their chances of getting something back.

The Need for Variability

For any kind of innovation to occur, there must be variability somewhere, either in the lifetime of an individual or in evolutionary time. We discussed how this might arise and agreed that a totally stereotyped animal following only one rule could be trapped like an insect against a window. To get out it would either need some noise in its system or more than one rule.

ANIMAL FEELINGS

Having spent most of our time assiduously avoiding the subjective aspects of animal minds, we at last came to face the problem of animal "suffering," surely a subjective term if ever there were one. Even here, however, we found it most profitable to keep to observable behavior. We had previously discussed the idea of animals showing a capacity to evaluate their environments. Now we applied this in the context of animal feelings about their environments (5). Animals can choose between two environments, and in this sense they can tell us which they find preferable. But this in itself is an inadequate criterion of their evaluation: knowing that a person ranks smoked salmon more highly than paté tells us little about how much he likes or dislikes paté. However, animals can be asked how much they value certain commodities. Battery-caged hens, for example, prefer cages with litter or sawdust on the floor to wire cages. They can also be asked how much they are prepared to pay for access to litter. Do they find it "worth" going without food, for example? Will they "work" and take time to obtain it? This leads to an operational definition of suffering: animals suffer if they are kept in an environment that they have shown they will pay a great deal to get out of. We can make use of their own evaluation mechanisms to give us an animal-centered view of the world.

REFERENCES

(1) Abegglen, J.J. 1981. On Socialization in Hamadryas Baboons. East Brunswick: Associated University Presses, in press.

(2) Cowie, R.J.; Krebs, J.R.; and Sherry, D.F. 1981. Food hoarding by marsh tits. Anim. Behav., in press.

(3) Dale, R.H.I. 1981. Parallel-arm maze performance of sighted and blind rats: Spatial memory and maze structure. Behav. Anal. Lett.: in press.

(4) Dawkins, M. 1971. Perceptual changes in chicks: another look at the 'search image' concept. Anim. Behav. 19: 566-574.

(5) Dawkins, M. 1980. Animal Suffering: the Science of Animal Welfare. London: Chapman and Hall.

(6) Dawkins, R., and Krebs, J.R. 1979. Arms races between and within species. Proc. Roy. Soc. Lond. B 205: 489-511.

(7) Garcia, J.; Kimmeldorf, D.J.; and Koelling, R.A. 1955. Conditioned aversion to saccharin resulting from exposure to gamma radiation. Science 122: 158-159.

(8) Green, D.M., and Swets, J.A. 1966. Signal Detection Theory and Psychophysics. New York: Wiley.

(9) Harley, C. 1981. Learning the ESS. J. Theor. Biol.: in press.

(10) Herrnstein, R.J. 1970. On the law of effect. J. exp. Anal. Behav. 13: 243-266.

(11) Krebs, J.R.; Kacelnik, A.; and Taylor, P. 1978. Tests of optimal sampling by foraging great tits. Nature 275: 27-31.

(12) MacArthur, R.H., and Wilson, E.O. 1967. The Theory of Island Biogeography. Princeton: Princeton University Press.

(13) O'Keefe, J., and Nadel, L. 1979. Precis of O'Keefe and Nadel's "The hippocampus as a cognitive map." Behav. Brain Sci. 2: 487-533.

(14) Olton, D.S., and Samuelson, R.J. 1976. Remembrance of places passed: spatial memory in rats. J. Exp. Psychol. Anim. Behav. Proc. 2: 97-116.

(15) Olton, D.S.; Becker, J.T.; and Handelmann, G.E. 1979. Hippocampus, space, and memory. Behav. Brain Sci. 2: 313-365.

(16) Packer, C. 1977. Reciprocal altruism in Papio anubis. Nature: 265, 441-443.

(17) Perrins, C.M., and Moss, D. 1975. Reproductive rates in the great tit. J. Anim. Ecol. 44: 695-706.

(18) Reither, F. 1980. Selfreflective cognitive processes: its characteristics and effects. Paper presented at the XXII International Congress of Psychology, Leipzig.

(19) Reither, F. 1981. About thinking and acting of experts in complex situations. Simul. Games, in press.

(20) Roberts, W.A., and Dale, R.H.I. 1981. Remembrance of places lasts: proactive inhibition and patterns of choice in rat spatial memory. Learn. Motiv.: in press.

(21) Rothstein, S.I. 1975. Evolutionary rates and host defenses against brood parasites. Am. Natural. 109: 161-176.

(22) Rozin, P. 1976. The selection of food by rats, humans and other animals. In Advances in the Study of Behavior, eds. J.S. Rosenblatt, R.A. Hinde, E. Shaw and C. Beer, vol. 6, pp. 21-76. New York: Academic Press.

(23) Sherry, D.F.; Krebs, J.R.; and Cowie, R.J. 1981. Evidence for memory of food storage sites in the marsh tit. Anim. Behav.: in press.

(24) Thorpe, W.H. 1963. Learning and Instinct in Animals. London: Methuen.

(25) Tinbergen, L. 1960. The natural control of insects in pine woods. I. Factors influencing the intensity of predation in song birds. Arch. Neérl. Zool. 13: 165-343.

(26) Tolman, E.C. 1948. Cognitive maps in rats and men. Psychol. Rev. 55: 189-208.

(27) Trivers, R.L. 1971. The evolution of reciprocal altruism. Q. Rev. Biol. 46: 35-57.

(28) Uexkull, J. von. 1934. **Streifzüge durch die Umwelten von Tieren und Menschen. Berlin: Springer.** Translated in Instinctive Behavior, 1957, ed. C.H. Schiller. London: Methuen.

(29) Welker, C. 1981. Natural and dependent rank of female crab-eating monkeys (Macaca fascisularis) in captivity. In Proceedings of the International Primate Society. Berlin: Springer, in press.

(30) Wiley, R.H., and Hartnett, S.A. 1980. Mechanisms of spacing in groups of juncos: measurement of behavioural tendencies in social situations. Anim. Behav. 28: 1005-1016.

Group on
Comparative Approaches to Animal Cognition

Standing, left to right: John Cerella, Friedrich Wilhelm Hesse, Rüdiger Wehner, Henry Gleitman, Richard Brown, Jerry Fodor, John Crook, Gerd Lüer, Paul Rozin.
Seated: Walter Kintsch, Douglas Gillan, Friedhart Klix, Don Griffin, Rainer Kluwe.

Animal Mind - Human Mind, ed. D.R. Griffin, pp. 375-389.
Dahlem Konferenzen 1982. Berlin, Heidelberg, New York: Springer-Verlag.

Comparative Approaches to Animal Cognition State of the Art Report

W. Kintsch, Rapporteur
R. H. Brown, J. Cerella, J. H. Crook, J. A. Fodor,
D. J. Gillan, H. Gleitman, D. R. Griffin, F. W. Hesse,
F. Klix, R. H. Kluwe, G. Lüer, P. Rozin, R. Wehner

INTRODUCTION

Our discussions of animal cognition were strongly influenced by the composition of our group, which ranged over a very broad spectrum from ethology to operant conditioning and cognitive psychology to social psychology. To find some common denominator in that diversity proved to be an interesting and stimulating task. What eventually emerged was a rephrasing of our initial discussion topic, and a concentration on the potential contributions that human cognitive psychology could make to the study of animal cognition.

Before summarizing these discussions, some general points need to be made. First, we shall, for the sake of convenience, continue using the term "animal cognition," but want it understood that we do not thereby deny cognitive differences among species. Much of the work we discussed involved chimpanzees because they are close relatives and more data are available than for other species.

Throughout our discussions there was an awareness of the importance of evaluating animal cognition in a natural ecological situation. This does not, however, imply a rejection of laboratory experiments. It is one question to determine an animal's peak intellectual functioning in its natural niche, another to explore the extent to which it is limited to that niche.

Cognition is, of course, itself a complex term, and it must be understood that its interesting aspects can vary among several independent dimensions. Computational complexity is one dimension, but the plasticity of a behavior and its accessibility - the extent to which it is situation bound - may be equally important.

Finally, we want to stress that by asking what human cognitive psychology can do for the study of animal cognition, we do not want to imply that it has all the answers, just for the asking. Often, the problems that cognitive psychologists have worked on successfully are very narrow ones. Only in relatively recent years has there been a determined movement away from the study of simplified laboratory problems to complex real world tasks (e.g., Lüer, this volume). For many interesting problems psychologists still have nothing to say, e.g., the problem of consciousness that continues to fascinate students of animal cognition for obvious reasons. Other crucial areas where modern cognitive theory must pass are motivation and learning. However, even where there is nothing positive, a history of unsuccessful approaches can in itself be helpful sometimes.

INTERNAL REPRESENTATION

A discussion of animal cognition does not necessarily imply a theoretical commitment. One could attempt to explain the phenomena of animal cognition in behavioristic terms. However, this was not the approach taken here. A cognitive theory of animal cognition seemed more appropriate. Therefore, we start by examining some basic properties of such theories, in the hope of thereby clarifying the actual task of theory construction (for general references see (1,5,6)).

A central concept in a cognitive theory is that of representation (see Cooper, this volume, for a discussion of this concept). It is important to be clear about what is involved in a representational theory of the animal mind, because our research strategy will, in part, be determined by the nature of such a theory. The view we take here is of the organism (human, animal) as a computational, inference making system - starting with perceptual and motor integration and continuing to reasoning and problem solving. This view of the organism is not new. The basic model goes back as far as Locke and Hume, and with it the tactics of theory construction; only the tools for theory building are new (mathematical systems, computer systems).

Mental representations are invented by theorists to permit the modelling of operations in a belief system. Beliefs have peculiar logical properties: both causal ones (a belief causes behavior, a sensation causes a belief) and semantic ones (beliefs are true or false, they are about something). (Note that belief is used here in a technical philosophical sense, not in its everyday meaning.) Mental symbols - representations - were postulated to reflect these properties. Physical symbols have this dual character (the printed symbol DOG on this page both causes a sensation when I read it and also refers to dogs). Mental symbols are just like that (e.g., Hume's "Ideas"): they have causal properties (the laws of association were at one time thought to be an adequate model thereof), but at the same time they are images (they have semantic properties, they refer to something).

Inferences are causal relationships among beliefs. Thus, my wanting to do something, causes the action; my knowledge about some state of affairs causes a corresponding statement. Inferences are strategies for distributing truth from one belief to another: given some beliefs (premises), causal relationships give rise to new ones (conclusions), either with logical necessity or merely with probabilistic certainty.

This general philosophical argument was presented here because of its implications for research strategy. It implies that the proper task of students of animal cognition who desire to formulate a representational theory of the animal mind is a complex one. First, a survey of the cognitive capacities of organisms must be made, then a formal system must be constructed that can account for these observations. The second half of the task cannot be approached until an extensive and systematic observational base is available to us.

A research strategy is of course not a recipe for doing good research. Even if one agrees on the general nature of the problem, there are difficult and tricky questions ahead. A basic and crucial one has to do with the following question: which of the endless number of observations on animal cognition are the theoretically significant ones, the ones we should really take seriously? Since we have as yet no theory of animal cognition, we cannot tell - but as scientists we have to make a commitment, a best guess, as to what we think will pay off in the future. A second major problem arises when we approach the task of formalization because of the peculiar interdependence of representation systems and process models. For our theory we need to have both. The problem is that a given set of observations can in principle be accounted for by many different representational assumptions coupled with different processing assumptions: an animal may have Cognitive Map A and use it in the manner described by Model A', or it may have the Cognitive Map B and the operation of Model B' - both representation/process pairs can result in an identical set of observations.

But to have a clear conception of the nature of the enterprises and its complexity does not imply giving up in despair. On the contrary, the task of formulating a cognitive model of animal behavior is being vigorously pursued today by many investigators and deservedly so. The discussion of the group did not center upon the success so far achieved. (As a prototype for a successful cognitive model in this area, consider Sutherland's work on the octopus: a mapping out of the visual abilities of the animals, followed by the construction of a representational system and a process model that described the observed discrimination behavior (14).) Instead, we concentrated on future research: what can, should be done next? What are the most promising problems? The most important questions? More specifically, given the mix between biologists and psychologists in our group, what is there in human cognitive psychology that could be used to guide research in animal cognition? Are there certain ways to state the questions that might be useful in the animal area? Are there techniques that could be adapted?

The group's discussion on these problems ranged over a very broad area. In outlining some of its main points we begin with "simple" cognitive processes, such as perception and motor action, and continue through memory, learning, and decision making, to the more esoteric problems of metacognition, self-concept, and consciousness which are especially interesting in the context of animal behavior.

INTEGRATIVE PROCESSES IN ACTION AND PERCEPTION

It is widely believed today that human perception is a constructive process, that it involves building a model of the world. The investigation of inferential processes in animal perception would appear especially promising, because the ecological situation for at least some animals and man is not very different. At the same time, this is the least mental part of cognition, and problems arise in the animal - man comparison. At what levels should they be compared: obviously not at that of particle physics, or at the neuronal level. The perceptual system functions as a hierarchy of subroutines, and one of the interesting problems is at what level of this hierarchy humans and animal(s) begin to diverge. It seems likely that in many animals alternative forms of dealing with the perceptual world have evolved which show few interesting communalities with human perception.

Differences may exist in sensory modalities (electroreception, for instance), or in motor capabilities (such as flying versus walking). Even so, imaginative speculation, later constrained by data gathered for this purpose, might be able to bridge the gap between animal and human world. (Griffin: "Contra Wittgenstein, if a lion could talk I suspect we could understand a significant fraction of what he might tell us.")

The studies of "perceptual intelligence," which are proposed below, still need considerable improvements in their theoretical underpinnings. It simply will not do to use such global, undifferentiated terms as "perception" or "intelligence." We have to define what we mean in operational as well as theoretical terms with perception: recognition? encoding? decision making?

Many discussions about perception center around the concept of complexity, which is, however, intolerably fuzzy. We need objective measures of complexity which are widely usable (the ones we have, such as information in bits, various indices of form complexity are not generally useful). Only if we have such measures could we answer questions whether complexity is perceived in the same way by humans and animals, or differently by animals.

Nevertheless, there are a number of investigations of inferential processes in perception that would appear to be worthwhile. One cluster of studies deals with the questions about whatever indications are available, or could be obtained, to reveal how animals construct a model of their environment. For instance, animal research on apparent motion might tell us something about the role of inferences in motion perception. Cues such as perspective, occlusion, and other depth indicators could similarly be explored. The temporal accumulation of information about the world could be studied in experiments where animals are permitted only brief glances at the environment, or where information is presented separately in different sensory modalities.

A somewhat different type of question concerns natural categories in different animals. What objects do animals perceive as belonging to the same category? Herrnstein-type experiments could be expanded to study a variety of concepts in the pigeon and other species. Could animals learn to distinguish among events or things that are not part of their natural environments (4)? Could they discriminate the temporal sequence of past events? A basic question here is the sense in which "events" have the same meaning for animals and men. Do animals segment time in units such as "searched for food," "returned to the nest," etc. (our equivalents of "went to the movies" then "had ice cream")? Conceivably, there are other, more micro segmentations that might comprise the "natural" level of animals (on the concept of a natural level see (12)). Could they discriminate earlier versus later in the more or less recent past? Oddity discrimination might provide another method to investigate such questions, if not in the pigeon then perhaps in monkeys.

Parallel to perceptual integration there is a set of questions concerning motor integration and the planning of actions. Detour and blocking paradigms have been used successfully in previous investigations of such problems. A more detailed study of action planning in animals would appear especially promising because of the widely held view that action is ontogenetically a precursor of thought. The child and presumably the animal first represents the world by acting on it, or interacting with it, before thinking about it (9).

An important problem here concerns the issue of automatization of learned behavior. Perceptual-motor behavior, just like other behavior, is first slow, requires conscious, undivided attention (a high resource demand in the terminology of cognitive psychology), but may with continued practice become automated: fast, unaware, open to time sharing (reading, typing, and driving a car are familiar examples, e.g., (13)). In order to understand behavior we must study it both in its automatized form and in its transition to automaticity. Concepts and methods from cognitive psychology (dual task designs, models of attention, resource-performance functions, utility analyses, e.g., (7,8)) are readily available to the student of animal cognition.

ELEMENTARY INFORMATION PROCESSING MECHANISMS

Cognitive psychology has had considerable luck with the concept of information processing components beyond the perceptual level, in such areas as memory and problem solving. Such concepts as a limited short-term memory buffer, retrieval operations in memory, semantic structure, planning strategies, etc., are involved. Obviously, we need to devise memory representations that are compatible with perceptual representations. Various pattern matching devices can be distinguished that operate on these representations (comparison of properties, template matches). Models for long-term memory (semantic memory, general knowledge) need to be expressed in a prelinguistic, conceptual form. As was already mentioned, each representational system implies its own processing assumptions.

It would be of considerable interest to study systematically the elementary information processing mechanisms in different species. Studies of short-term memory capacity and divided attention in automatized and non-automatized behaviors could probably be designed. Similarly, one probably could (at least for the chimpanzee) answer the question whether search times in short-term memory as well as matching times for both physical identity matches and other (e.g., cross-modal) matches are comparable to the human case. With the close identification of short-term memory and human consciousness, this might also be one way to get indirectly at the more complex questions to be raised later.

Suppose one finds differences in the short-term memory capacity among the apes; might these then be related to differences in their capacity to tolerate delays in a response prior to action, which appears to be a precondition for much intelligent behavior? Might these differences furthermore be related to the ecological situation of the different apes, are they consequences of living in different environments?

One motivation for such research on information processing mechanisms in animals would be that it could provide a sound basis for other research that is of interest to students of animal cognition. Once we have such knowledge about an animal, the search for electrophysiological correlates of cognition (not only evoked potentials but also endogenous ones) could be guided better from the behavioral side. Similarly, the ethologists' study of tool use in animals could advance further. The relation of tool use to cognitive capacities could be determined, and a more detailed description of the actual processes and their underlying mechanisms involved in tool use could be attempted. One can doubt that it is possible to answer the interesting and important question as to the role of culture in tool use definitively - unless perhaps, one gets a clear idea of what the mechanisms involved are.

ON INTELLIGENCE AND PROBLEM SOLVING

Within psychology, the information processing view has largely replaced such global concepts as "intelligence" and "general problem solving." Older views (intelligence is what the test tests, the multidimensional, multifactor theories of intelligence) have given way to concerns about the mechanisms involved in intelligent behavior and individual differences in the nature and use of these mechanisms. There is no point in asking what intelligence is (a variation of this argument applies to problem solving), but rather what we need is to match a detailed task analysis of some intelligent behavior to a detailed analysis of an individual's information processing capacities (e.g., (11)), and to study the fit between the natural environment and specific "intelligent" behaviors.

At least some group members felt that this shift in research emphasis has important consequences for the study of animal cognition. Evolutionary biologists often believe that even something as diffuse as intelligence might well be selected for in evolution if it was strongly influenced by genetic factors. There is, however, a counterposition which might be worth studying. Adaptation and selection are of primary interest to the biologist, but we have to ask ourselves with great care: selection for what? The answer advocated here is that there can be no selection for intelligence (or creativity, or problem solving ability), because these are not unitary traits that characterize humans or animals. The selection must operate on other, more basic properties that are linked to what we call "intelligence." It becomes crucial then to understand what these properties are. One cannot understand the evolution of intelligent behaviors if one looks for the evolution of "intelligence," because that comprises an extremely diverse and task dependent assortment of information processing mechanisms. Instead, these mechanisms themselves deserve our attention. As well, the extent to which any intelligent system is limited to specific input or output channels, or motivational states, and the process through which, in development or evolution, limited systems become more generally available is worthy of study.

It is in no way argued here that we should not (or cannot) study what in everyday language is called intelligence. Only that we go about it the wrong way if we do it in the traditional familiar way.

METACOGNITION, CREATIVITY, AND AWARENESS

Here we come to the concepts that are of primary interest to students of animal cognition. The relationship between megacognition (knowledge about one's own cognitive processes, see Kluwe, this volume) and consciousness is a complex one: in particular, metacognition might exist without consciousness. While the group felt unable to define "consciousness" (Fodor: "You can milk a cow without defining it"), a great deal of the discussion centered around the intriguing question of how we can tell that animals are conscious. There was considerable interest in current research on metacognition with human subjects, and some suggestions as to how such research could fruitfully be extended to animals. Do wolves hunting cooperatively in a pack (or lions!) know what they are doing? Are they planning their hunt, or is it all just a lucky accident (or genetically preprogrammed)? An animal showing surprise indicates that it was expecting something that did not happen, but to what extent is awareness involved? Are sudden insights in animals (which are often reported in the literature) evidence for awareness? Perhaps, the experimental procedures of Premack and Woodruff can be adapted to investigate such questions (for a discussion of these issues, see Gillan, this volume).

Other questions of the same kind appear farther from solution at present, e.g., whether animals deliberately and without any immediate consequences teach other animals, or, the extent to which animal cultures may amplify the problem solving capacities of their members (2). With problems like these, clearly the first step is to collect relevant anecdotal observations.

Some group members took it for granted that animals must have a self-concept. For others, this appeared to be an empirical question. The work of Savage-Rumbaugh (see Seyfarth et al., this volume) appears to lend itself to an investigation of that question. A number of experimental designs were suggested that could, in principle, demonstrate that chimps have a concept of "self." By using video recordings, situations can be arranged in which the chimp can show that he is reacting to his own picture - not just some particularly entertaining image.

Creativity in animals was also extensively discussed. One experimental paradigm of great promise in this area is that of Karen Pryor, who trained porpoises to do a new trick every day (10). Can other animals learn similar tasks? How widespread is such behavior in the natural environment? While training is probably restricted to pretty smart creatures, it was generally agreed that even for the honeybee an experiment could be designed to investigate creativity. Bee swarming is a unique event in the life of an individual bee. The behavior involved is presumably genetically preprogrammed - or is it an instance of creative problem solving? What is needed is to arrange some other unique event in the life of bees - e.g., a sudden need to employ some particular kind of building material never before used by them. If the scouts dance to advertise the location of the new material, and if the workers understand them, do we have an instance of creativity in bees? It seems almost certain that imaginative investigators will answer many of the questions that were raised here - or tell us why they could not be answered in the form we have raised them.

CONCLUSION

Much of the dynamics of the discussion in our group was dominated by the differences in the world view of biologists and cognitive psychologists. Methodologically this was reflected in preferences for naturalistic observation versus laboratory controls. For some questions, it makes no sense to study animals without regard of their natural environment, and good laboratory studies must take this into account. Their function is not to replace naturalistic observation and anecdotal report, but to supplement them. The anecdote is the crucial first step toward a scientific theory, but it need not be the last one. Clearly, observations of rare, not easily replicable events play a fundamental role in the study of animal behavior. But there are possibilities to make these rare events more easily observable, under at least partially controlled conditions - e.g., by using a semi-naturalistic laboratory environment, as in some of the work with chimpanzees. It is a captive situation intermediate between a field situation and a completely structured laboratory situation, and probably as good (though expensive) a compromise as is possible today. It is certainly one of the greatest contributions that psychology can make right now to the study of animal behavior: psychologists know about the importance of instructions, they can teach others how to ask animals to talk to us. The work of the Premack laboratory (see Gillan, this volume) is exemplary in this respect.

By orienting themselves on the current state of the art in the more fully explored field of human cognition, students of animal cognition can at least avoid entering the same series of blind alleys that their counterparts had to pass through. There is always a tendency to do exactly that. There is no point in resurrecting, for the study of animals, methods and theories that have failed in the past for the study of human cognition: this holds true equally for theoretical constructs such as intelligence (11), theories such as the motivational theory of Hull (3), or methods such as the use of multidimensional scaling techniques and factor analysis as induction procedures. Students of animal behavior can learn from the failures of their colleagues, too.

Does a comparative approach to human and animal cognition have anything to offer for the understanding of the animal mind? On the basis of our group discussion, one would answer this question with a clear yes.

REFERENCES

(1) Anderson, J.R. 1980. Cognitive Psychology and Its Applications. San Francisco: Freeman.

(2) Borner, J. 1980. Animal Culture. Princeton: Princeton University Press.

(3) Bower, G.H., and Hilgard, E.R. 1980. Theories of Learning. Englewood Cliffs, NJ: Prentice-Hall.

(4) Herrnstein, R.J., and de Villiers, P.A. 1980. Fish as a natural category for people and pigeons. In The Psychology of Learning and Motivation, ed. G.H. Bower, vol. 14. New York: Academic Press.

(5) Klix, F., and Hoffman, J. 1980. Cognition and Memory. Berlin: VEB Deutscher Verlag der Wissenschaften.

(6) Kintsch, W. 1977. Memory and Cognition. New York: Wiley.

(7) Navon, D., and Gopher, D. 1979. On the economy of the human processing system. Psychol. Rev. 86: 214-255.

(8) Norman, D.A., and Bobrow, D.J. 1975. On data-limited and resource-limited processes. Cog. Psychol. 7: 44-64.

(9) Piaget, J. 1937. La construction du réel chez l'enfant. Neuchâtel: Delachaux et Niestlé.

(10) Pryor, K.W.; Haag, R.; and O'Reilly, J. 1969. The creative porpoise. Training for novel behavior. J. Exp. Anal. Behav. 12: 653-661.

(11) Resnick, L.B. 1976. The Nature of Human Intelligence. Hillsdale, NJ: Erlbaum.

(12) Rosch, E.; Mervis, C.B.; Gray, W.; Johnson, D.; and Boyes-Braem, P. 1976. Basic objects in natural categories. Cog. Psychol. 8: 382-439.

(13) Shiffrin, R.M., and Schneider, W. 1977. Controlled and automatic human information processing II: Perceptual learning, automatic attending, and a general theory. Psychol. Rev. 84: 127-190.

(14) Sutherland, N.S. 1959. Stimulus analyzing mechanisms. In Proceedings of a Symposium for the Mechanization of Thought Processes, vol. II. London: H.M. Stationary Office.

Group on
Communication As Evidence of Thinking

Standing, left to right: Jim Gould, Bob Solomon, Herbert Terrace, Martin Lindauer, Colin Beer, Dan Dennett.
Seated: Bob Seyfarth, Sue Savage-Rumbaugh, Carolyn Ristau, Peter Marler.

Animal Mind - Human Mind, ed. D.R. Griffin, pp. 391-406.
Dahlem Konferenzen 1982. Berlin, Heidelberg, New York: Springer-Verlag.

Communication As Evidence of Thinking
State of the Art Report

R. M. Seyfarth, Rapporteur
C. G. Beer, D. C. Dennett, J. L. Gould, M. Lindauer,
P. R. Marler, C. A. Ristau, E. S. Savage-Rumbaugh,
R. C. Solomon, H. S. Terrace

INTRODUCTION

This report examines ways in which studies of animal communication can increase our understanding of animal minds. The report begins by offering working definitions of two particularly difficult concepts, consciousness and intentionality. Second, the report examines whether the use of such definitions helps to understand the mental events that may be occurring during particular communicative acts. Referential communication, deceit, and reciprocal altruism are considered as examples. Third, the report discusses the analysis of shared conventions and internal representations in a variety of animal signals. Finally, the report considers certain findings from the ape language projects which seem to offer new insights into the problem of animal minds. Discussion emphasizes the relation between this work and studies of natural communication.

CONSCIOUS VERSUS UNCONSCIOUS "THOUGHT"

Grasping the nettle that others have chosen to avoid, our group began by attempting to define the difference between conscious and unconscious thought. For me, the most illuminating point to emerge from this exercise was that we should not set our expectations too high. If we aim for a rigorous definition that offers a widely applicable characterization of different behaviors as "conscious" or "unconscious," we are likely to fail. Such failure would occur largely because we do not yet fully understand what we are attempting to define. Moreover, the notion that across the animal kingdom there are abrupt discontinuities between species in levels of consciousness is almost certainly false.

On the other hand, as Gould has argued, if we aim for a working definition that may be formally incomplete but nevertheless focuses our attention on certain issues, this may in turn lead us to design better experiments, and we are more likely to succeed in our ultimate goal. In other words, the essential feature of our definition is that it should be both intuitively satisfying and operationally useful.

Gould describes a necessary condition for consciousness as "an ability to recognize, in a variety of cases, a logical conflict between a token or sign stimulus and the context which it is supposed to represent." Consider the case Gould describes of a bee using the odor of oleic acid as a token to indicate a dead bee, whereupon it removes the corpse from the hive (27). By the operational definition described above, we can provide a logical mismatch - a live bee with a drop of oleic acid on it - and observe that other bees nevertheless (and with considerable difficulty) throw the bee with acid on it out of the hive. Clearly the bees cannot be said by this definition to be in any meaningful sense conscious of their actions. An equivalent approach can be applied to supposedly higher intellectual functions.

Note that the approach taken above does not yield a definition in the proper sense, for three reasons: first, it provides only a necessary but not a sufficient condition for consciousness; second, it does not touch on experimentally unapproachable questions of subjective awareness; and third, it neglects the central question of our endeavor: what is the <u>mechanism</u> by which recognition of a mismatch takes place? Is the behavior invariant and hard-wired, demonstrating the elegance of evolution, or is it flexible in one or more ways that suggest the elegance of cognition? The definition above does not provide an answer, but it does pose the question in a way that may be useful in dealing with specific examples (see below).

<u>INTENTIONALITY</u>

Dennett (3) has offered a philosophical analysis of what we might call "mental states" ina form that may be useful to those studying animal communication. His formal argument is summarized only briefly here. In following sections his reasoning is applied to specific problems in the analysis of behavior, with what we believe are useful results.

Dennett begins by arguing that within the animal kingdom there is a continuum of what we may call mental states, from the simplist self-regulatory systems to our own mind. Within this continuum, it seems unlikely that there will be abrupt discontinuities. Thus our aim (as with consciousness - see above) should not be to call some species mental and others not. Instead, we should set out to examine degrees of complexity that are exhibited in different species. This requires some method of quantification, and Dennett proposes that we begin by drawing on our own descriptions of behavior and ascribing some <u>hypothetical</u> level of intentionality to the animals we are observing. Does one squirrel threaten another "because it is excited" (what Dennett would call zero order intentionality)? Does the squirrel threaten because it <u>wants</u> its opponent to leave the area (first order intentionality)? Or does the squirrel threaten because it <u>wants</u> its opponent to <u>believe</u> that it will attack (second order)?

Note that, as this analysis begins, the act of ascribing a hypothetical level of intentionality does not really tell us anything. It does, however, offer a means by which we may compare different levels of mental complexity and devise experiments that allow us to choose one level over another. As explicated below, the basic strategy is to describe a behavior in the most complex intentional terms that seem reasonable and then to perform certain logical operations (or substitutions) which serve to strip away the irrelevant complexity and eventually describe the behavior in the simplest possible terms, that is, the fewest intentional steps. As will become clear (see below), this is basically a more formal, intellectually elegant version of the test implied in Gould's working definition.

HOW USEFUL ARE THESE CONCEPTS? SOME REAL-LIFE EXAMPLES

Referential Signals

Consider the following example. A group of East African vervet monkeys is foraging on the ground in an area of open woodland. Suddenly, one monkey sees a leopard and gives an alarm call. Without a second's hesitation the others run up into the nearest tree. If the alarmist had seen an avian predator, it would have given an acoustically different alarm call, and its fellow group members would have responded by looking up into the air or running into bushes. The sight of a python would have produced a third alarm call, to which the others would have responded by looking down on the ground around them (17,18,21).

What are the important questions to be asked about this system of signals, and how do our working definitions of consciousness and intentionality help us answer them? One starting point might be to investigate possible levels of complexity in the mental state of the monkey who gives the alarm: what information does he intend to convey? Using Dennett's formulation, we may begin with the simplest hypothesis: the monkey gives an alarm because it is excited (zero order intentionality). This we may tentatively discount because of the observation that lone monkeys who encounter predators do not give alarm calls. (Of course, there are other possibilities: no exhaustive hypothesis testing is implied here. I simply introduce the manner in which an explication might proceed.) Second, one may hypothesize that the monkey gives an alarm because it <u>wants</u> the others to run into a tree (first order intentionality). The observation that monkeys will alarm-call at leopards even if all the members of their group are already in trees suggests that this explanation can also be eliminated. Third, we may hypothesize that the monkey gives an alarm because it _wants_ others to _believe_ that there is either (a) something interesting nearby, (b) a predator nearby, or (c) a leopard nearby (all second order). To date, field observations suggest that explanations (a) and (b) can be eliminated, leaving explanation (c) as, thus far, the most strongly supported by data (17,18). In this case, Dennett's method allows us to examine preexisting data, perhaps yielding a better understanding of just what is occurring when a vervet monkey gives an alarm call. It is hoped that, had we not had such observations already in hand, either Gould's or Dennett's analytical scheme would have suggested specific experiments (see also below).

Note that in the course of this analysis we have at least some opportunity to deal directly with the specificity of meaning in animal signals. Posing the question "What is it that the monkey wants?" generates at least some hypotheses about the monkey's internal representation that can be tested further in the field (see also below). What this particular analysis does _not_ tell us is just how much the monkey knows about the relation between signal and referent (16), or about the cause-effect relation between signal and response. Such questions, of course, are not meant to apply only to monkeys, since predator-specific (or at least predator class-specific) alarm calls are widely known in many insects, birds, and mammals (e.g., (14,19)). Could an insect, bird, squirrel, or monkey - even in the absence of a predator - make use of its knowledge that certain alarms elicit certain responses? The question is particularly intriguing because it leads directly into one of the knottier problems in ethology: the notion that animals play tricks on each other.

Deceit

Cases of communicative deceit are well-known throughout the

phylogenetic continuum and offer a special fascination for the devious minds of our species. Ants, for example, have two specific alarm signals - olfactory signals in this case - one which directs attack at the predator while the other elicits fleeing (26). Certain species of slave-making ants regularly begin their raids on other colonies by spraying the "run away" odor to empty the nest before beginning their attack. Using our working definitions as a starting point, we can formulate tests to elucidate the level of intentionality exhibited in such behavior, and we may ultimately arrive at a quantitative comparison between this sort of deceit and other, perhaps more complex, varieties.

The data available on other cases of deceit, however, are not yet sufficient for such an analysis. For example, if a lion approaches an ostrich that is guarding its nestlings, the adult bird will run some meters from the nest, fall to the ground, and give what is most vividly described as a "broken wing display." To what degree of mental complexity should we attribute this behavior, which occurs in many avian species? Initially, we can probably eliminate the hypothesis that the bird displays simply because it is frightened or excited. Similarly, experiments could be devised to test whether the ostrich wants the lion to run away, or (still more complex) wants the lion to believe that it is injured. Intriguingly, there are a number of reports in the literature indicating that in some avian species the response of the parent depends a great deal on the nature and behavior of the potential predator. Kildeer plovers, for example, respond to the approach of a dog or human with a distraction display that leads the predator away from the bird's nest. In contrast, when a cow approaches, the bird remains near its nest, flying conspicuously in the cow's face (23). Just how modifiable is the display given various additional external stimuli? What would a bird do if a predator were approaching its nest and yet there were some preferred prey (such as a real wounded animal) nearby?

An even more Byzantine problem comes from work by Woodruff and Premack (28) on chimpanzees. In their experiments a caged chimpanzee watched food be hidden in one of two containers. After a short time interval, a human trainer entered the room. If the chimpanzee indicated to the trainer which container held food, the trainer would then share the food. On other occasions, however, a second trainer might enter the room after food had been hidden. Now if the chimpanzee indicated where the food was located, the trainer ate it all himself.

As Woodruff and Premack's experiment progressed, the chimpanzee became increasingly uncooperative toward the second trainer, first by withholding information and second by actually indicating the incorrect container. What can we say about the chimp's behavior in this case?

The first question to ask is: why is the chimp not puzzled when the bad trainer, having gone to the incorrect box, fails to go to the correct one? The chimp's lack of puzzlement (if it exists) suggests that he either (a) does not really understand the relation between the trainer's knowledge and his actions, or (b) believes the trainer is exceedingly stupid.

Second, on the basis of Woodruff and Premack's evidence, can we say that the chimpanzee wanted the trainer to believe that the food was in a particular container? Here Dennett and Savage-Rumbaugh offer a crucial experiment: place the food in plastic, see-through containers and instruct the bad trainer to look at these containers as he enters the room. If the chimpanzee can instantly recognize a logical inconsistency between (a) what it wants the trainer to believe and (b) what the trainer obviously knows, then we would have some stronger evidence for the notion of conscious deceit.

Reciprocal Altruism

As a final example of the potential for learning more about animal minds through studies of animal communication, consider the case of reciprocal altruism (25). Reciprocal altruism occurs when, over time, two individuals exchange altruistic acts. During the time between acts, the most recent altruist is thought of as "gambling" that his behavior will be reciprocated. Trivers (25) argues that reciprocal altruism can evolve if the cost of the altruistic act to the donor is less than the benefit which such behavior brings to another, possibly unrelated, individual (2,25). Thus far, the only well-documented case of reciprocal altruism in non-human species has been found among baboons (12). For the purposes of this report, reciprocal altruism is interesting because it raises the issue of deceit, and because it can usefully be analyzed using the theoretical framework outlined in earlier sections.

Grice (6) argues that language, and the cognitive abilities it implies, requires at least third order intentionality (for example, A wants B to recognize that A believes p). From an entirely different perspective, Trivers (25) has argued that reciprocal altruism, and the detection of cheaters that it implies, is responsible for the rapid evolution of cognitive capacities exhibited by the higher primates, especially man.

Thus if it can be shown that reciprocal altruism (including detection) is unique among bits of behavior in requiring a certain order of intentionality, there will be some support for Trivers' thesis. In fact, Crook (2) does make the point (though not in these terms) that reciprocally altruistic behavior requires at least third order intentionality: in our hypothetical view, the recipient wants the altruist to believe that he intends to reciprocate. Moreover, the detection of cheating, even in its simplest form, requires a higher level of intentionality than the original altruistic act, and increasing abilities to detect cheating involve increasing levels of intentionality.

Two points emerge from this discussion. First, it seems clear that there are striking parallels between the analysis of mental complexity by some philosophers and the analysis of reciprocal altruism in animal behavior. Is it in fact true that reciprocal altruism is the only sort of animal communication in which third order intentionality seems likely? Second, as Crook pointed out during discussion, higher orders of intentionality seem to necessitate the conceptualization of self as agent. Beer and Terrace also suggested that complex social behaviors provide the most likely setting for the emergence of self-awareness.

In summary, all of the examples cited above suggest the usefulness of thinking about consciousness in terms of a preliminary, working definition. Even if it is not possible to define such terms with formal precision, we can describe them in such a way that they lead to better questions and more useful experiments.

SHARED CONVENTIONS AND INTERNAL REPRESENTATIONS

Shared Conventions

In a variety of areas within animal communication, the analysis of shared conventions and internal representations can potentially provide deep insights into the nature of animal thought. We are thinking here of conventions shared both between animals and, perhaps more importantly, between different operations within the same animal.

Although shared conventions require common rules (or common neural substrates - see below), they do not in themselves require the invocation of internal representations on the part of the communicating animals. Social amoeba, for instance, are able to receive and transmit social messages (e.g., cyclic AMP as an aggregation signal, and surface chemicals for species recognition) with, one imagines, little need for mental imagery. Similarly, in higher animals such as bees there is little need to invoke thought or imagery in the mutually understood tactile signals transmitted by one bee to elicit food sharing by another.

However, at a certain level of neural complexity - a complexity related more to the task at hand rather than to any inherent phylogenetic complexity - it seems more efficient for animals to map a complex communicative input onto the same processing array used to direct the output. There is some evidence, for example, that crickets engage in such a strategy since hybrid females prefer the songs of hybrid males (8). At the very least, this implies that the same genes are used to build the encoding and decoding arrays. How much simpler if the arrays are identical.

Honey bees, for example, clearly have maps (cognitive maps in the usual sense) of the world around the hive ((4), see also Gould and Gould, this volume). They can place themselves and their food sources on this map and use it to travel from one place to another. Is this map used to encode the dance message, and could it be that recruits use the same array to place the goal wherein they decode the message? More speculatively still, might such input/output co-mapping represent a pre-adaptation for the formulation of the sort of internal representations familiar in humans?

When we consider vertebrates - birds and their learned songs, for example - various lines of evidence point toward the heuristic value of postulating common mechanisms for song reception and production. Birds, after all, need to learn both how to sing and how the songs of others should sound. In the course of such development, innate constraints on the learning process might ensure communality in the ways in which internal representations develop in different individuals of the same species (5). This would greatly facilitate the sharing of sufficient rules for two-way communication to become possible. Moreover, the sweeping neural economy of co-mapping the input and output is self-evident. It would, for example, suggest an explanation for the ability of naive human infants to view a face and mimic its expression (5, 11). The crucial point here, though, is the greatly expanded associational capacity such a neural strategy would make possible. We suggest that it might be quite fruitful to look for this neural strategy in animals and to examine the cognitive capacities it makes possible, as embodied in such hypothetical constructs as "templates" and "schemata" (5). The notion of shared central mechanisms for perception and production underlies the motor theory of speech perception (9,22). Common brain mechanisms are postulated, developing in infancy through the conjoint operations of perceiving speech and of speaking. These then take part in the control of both kinds of operation in the mature organism.

Internal Representation

There is a long tradition in behavioral research which denies the need for any sort of internal representation in animals. In constrast to this view, Terrace and others argued throughout the conference that virtually all animals have some sort of internal representation (even bees, for example, have some sort of mental map of their surroundings), and therefore the important question is not whether such representations exist but precisely what form they take.

In this regard, Terrace pointed out that traditional behaviorists have concerned themselves simply with one kind of representation, namely, stimuli and response occurring in a linear manner. Accordingly, they saw no need to postulate other, possibly cognitive variables that might mediate the behavior in question. Now, for the first time, there is a need to recognize non-linear S-R models (e.g., (20)), and to entertain the notion that animal cognition can be studied with models derived and validated in research on human cognition (Kintsch et al., this volume).

Olton's work on the behavior of rats in a radial maze argues strongly for the view that animals have complex maps of their surroundings (for further discussion see Dawkins et al., this volume). Concept formation by pigeons (1,7) offers another example of work that is not easily explained by simple (or even complex) S-R sequences. Given that a pigeon can group together pictures of landscapes according to whether they have people in them, or trees in them, how are we to characterize the pigeon's knowledge? Is it appropriate to use the same terms to characterize a bee's knowledge of the area near its hive?

The notion of an internal representation is often dealt with in terms of a species' "natural categories." A number of songbirds can not only distinguish different song types within a given species but also can discriminate among different notes within a particular song type (10). Does this justify the conclusion that the birds' natural categories are hierarchical, such that three song notes, for example, a_1, a_2, a_3, are distinguished individually yet all are seen as belonging to the set, or song type, A? What sort of data would constitute evidence for true hierarchical organization, and what sorts of selection pressures might favor the ability to construct such second order relations?

THE PROBLEMS POSED BY APES

No discussion of the relation between communication and animal minds would be complete without some mention of the ape language projects. Since a thorough review would be neither possible nor desirable (see Ristau and Robbins, this volume), it was decided to stress a number of points which in some respects represent a reassessment of the goals and aims of those working in the field.

As a starting point, Terrace argued that we should replace the study of syntax with a more detailed investigation of the meaning of individual signs. This follows from the fact that, whether because of motivational or cognitive deficits, most of the signs made by the chimpanzee in Terrace's study were simply repetitions of signs made shortly before by the animal's trainer (24). As long as this occurs, multiple sign combinations are unlikely to provide meaningful evidence of syntax, and syntactical approaches to the study of meaning (at least in work on captive apes) are unlikely to be successful.

With regard to the properties of meaning in individual signs, Terrace distinguished between (a) a chimp's ability to respond differently when it perceives different stimuli, and (b) the ability to use a symbol as an abstract term to refer to the relationship between different objects. In Terrace's study, for example (24), the authors found it relatively easy to teach the chimpanzee Nim the distinction between "bottle on the table" and "bottle under the table." It was far more difficult, however, to teach Nim the abstract concept "on," such that he could choose correctly between, for example, "on" and "under" regardless of the particular stimuli present.

In contrast, Premack (13) has achieved clear success in teaching the chimpanzee Sarah to use tokens that represent the concepts "same" and "different." Savage-Rumbaugh et al. (16) present clear evidence that chimps can label lexigrams as either "food" or "tool" depending on the objects which these lexigrams represent.

Terrace and Savage-Rumbaugh suggested that attention be focused on the chimp's ability to conduct two cognitive operations simultaneously: can it use one set of tokens to refer to different objects, while at the same time using a different set of tokens to refer to the relations that exist between those objects? For example, in Savage-Rumbaugh's (16) study, the chimps Sherman and Austin learned to use one set of tokens to refer to different objects (wrench, stick, etc.). They then learned a second set of tokens ("food" and "tool") to refer to the functional relationships between items of the first set. This approach to the chimpanzee's single utterances, it was argued, is perhaps the most likely way to increase our understanding of the chimpanzee's mental processes.

Finally, underlying all of these problems is the issue of motivation. Consider, for example, Terrace's point that humans frequently communicate without any apparent attempt to get something. In contrast, to get a lone chimpanzee in a laboratory to say "red," you have to give some immediate reward. In between these two extremes, chimps living together in captivity will frequently communicate to each other without any immediate material reward, and it is a common observation among free-ranging primates that most vocalizations occur spontaneously, in relaxed social settings, and in circumstances where the vocalizer (apparently) does not expect to get anything from his listeners. With this in mind, Savage-Rumbaugh posed the question: Can apes use symbols to describe things that have no immediate functional value for them? For example, can they use "tickle" to describe what other individuals are doing, or is "tickle" limited to a request for a game they want to play?

These observations remind us that we may seriously underestimate the cognitive (or the linguistic) abilities of chimps if the circumstances in which we examine their behavior are too far removed from their natural habitat. In cases where a number of chimps are living together, and all have learned a human invented, referential form of signaling, under what conditions will this novel form of communication be used in preference to the animals' natural signs? If natural signs are always chosen, this could mean that either we have underestimated the communicative power of natural signals, or we have failed to provide the animals with problems that can only be solved using their new signaling system. Finally, although the ape language projects do in some cases face problems of subject motivation, it is worth re-emphasizing that they stand as a challenge to those working on field studies of vertebrate communication. How common is referential signaling under natural conditions? And under what circumstances might an individual capable of such behavior gain a selective advantage over others? Such questions bear directly on the issue of communication and animal minds, but as yet they are only beginning to be asked.

Acknowledgements. I thank all those who offered comments on an earlier version. J.L. Gould, P. Marler, and H.S. Terrace were particularly helpful in preparing the final manuscript.

REFERENCES

(1) Cerella, J. 1979. Visual classes and natural categories in the pigeon. J. Exp. Psych.: Human Percept. Perf. 5: 68-77.

(2) Crook, J.H. 1980. The Evolution of Human Consciousness. Oxford: Oxford University Press.

(3) Dennett, D. 1978. Brainstorms: Philosophical Essays on Mind and Psychology. Harvester: Bradford Books.

(4) Gould, J.L. 1981. Ethology: The Mechanisms and Evolution of Behavior. New York: W.W. Norton, in press.

(5) Green, S., and Marler, P. 1979. The analysis of animal communication. In Handbook of Behavioral Neurobiology: Social Behavior and Communication, eds. P. Marler and J.G. Vandenbergh, vol. 3, pp. 73-158. New York: Plenum Press.

(6) Grice, P. 1957. Meaning. Phil. Rev. 66: 377-388.

(7) Herrnstein, R.J.; Loveland, D.; and Cable, C. 1976. Natural concepts in pigeons. J. Exp. Psych.: Anim. Behav. Proc. 2: 285-311.

(8) Hoy, R.R.; Hahn, J.; and Paul, R.C. 1977. Hybrid cricket auditory behavior: evidence for genetic coupling in animal communication. Science 195: 82-84.

(9) Liberman, A.M.; Cooper, F.S.; Shankweiler, D.; and Studdert-Kennedy, M. 1967. Perception of the speech code. Psychol. Rev. 74: 431-461.

(10) Marler, P., and Peters, S. 1981. Birdsong and speech: evidence for special processing. In Perspectives on the Study of Speech, ed. P.D. Eimas, pp. 75-112. Hillsdale, NJ: Lawrence Erlbaum Associates.

(11) Meltzoff, A., and Moore, M.K. 1977. Imitation of facial and manual gestures by human neonates. Science 198: 75-78.

(12) Packer, C. 1977. Reciprocal altruism in Papio anubis. Nature 265: 441-443.

(13) Premack, D. 1976. Intelligence in Ape and Man. Hillsdale, NJ: Lawrence Erlbaum Associates.

(14) Ryden, O. 1978. Differential responsiveness of great tit nestlings, Parus major, to natural auditory stimuli. Z. Tierpsychol. 47: 236-253.

(15) Savage-Rumbaugh, E.S.; Rumbaugh, D.M.; and Boysen, S. 1978. Symbolic communication between two chimpanzees. Science 201: 641-644.

(16) Savage-Rumbaugh, E.S.; Rumbaugh, D.M.; Smith, S.T.; and Lawson, J. 1980. Reference - the linguistic essential. Science 210: 922-925.

(17) Seyfarth, R.M.; Cheney, D.L.; and Marler, P. 1980. Monkey responses to three different alarm calls: evidence for predator classification and semantic communication. Science 210: 801-803.

(18) Seyfarth, R.M.; Cheney, D.L.; and Marler, P. 1980. Vervet monkey alarm calls: semantic communication in a free-ranging primate. Anim. Behav. 28: 1070-1094.

(19) Sherman, P.W. 1977. Nepotism and the evolution of alarm calls. Science 147: 1246-1253.

(20) Straub, R.O.; Seidenberg, M.S.; Bever, T.G.; and Terrace, H.S. 1979. Serial learning in the pigeon. J. Exp. Anal. Behav. 32: 137-148.

(21) Struhsaker, T.T. 1967. Auditory communication among vervet monkeys (Cercopithecus aethiops). In Social Communication Among Primates, ed. S.A. Altmann, pp. 281-324. Chicago: University of Chicago Press.

(22) Studdert-Kennedy, M.; Liberman, A.M.; Harris, K.S.; and Cooper, F.S. 1970. The motor theory of speech perception: a reply to Lane's critical review. Psych. Rev. 77: 234-249.

(23) Taverner, P.A. 1936. Injury feigning by birds. Auk 53: 366-370.

(24) Terrace, H.S.; Petitto, L.A.; Sanders, R.J.; and Bever, T.G. 1979. Can an ape create a sentence? Science 200: 891-902.

(25) Trivers, R.L. 1971. The evolution of reciprocal altruism. Q. Rev. Biol. 46: 35-57.

(26) Wilson, E.O. 1971. The Insect Societies. Cambridge, MA: Belknap Press of Harvard University Press.

(27) Wilson, E.O.; Purlach, N.I.; and Roth, L.M. 1958. Chemical releasers of necrophoric behavior. Psyche 65: 108-114.

(28) Woodruff, G., and Premack, D. 1979. Intentional communication in the chimpanzee: the development of deception. Cognition 7: 333-362.

Animal Mind - Human Mind, ed. D.R. Griffin, pp. 407-414.
Dahlem Konferenzen 1982. Berlin, Heidelberg, New York: Springer-Verlag.

Afterthoughts on Animal Minds

T. H. Bullock
Neurobiology Unit, Scripps Institution of Oceanography and Dept. of Neurosciences, University of California, San Diego, School of Medicine, La Jolla, CA 92093, USA

"The evolutionary continuity of mental experience" is the subtitle of Donald Griffin's little classic of 1976 on "The question of animal awareness." The thrust of that subtitle and a major theme of the book is the proposition that the human species is not unique except in degree. That proposition is so consonant with our general biological understanding that there must be overwhelming agreement with Griffin, at least among biologists. However, there is still ample room for biologists to disagree on the matter of degree, on the matter of emphasis, and on the definition of mental experience. I, for one, regard the difference in degree to be so enormous that it is equivalent to a qualitative difference. That feeling was reinforced here by visiting the great Dahlem museum of art and the Berlin opera - and by our gathering here to address the issue.

In the background papers and in the group discussions of this workshop we have seen some of the disagreements, mildly and politely expressed as though they were purely academic. They become vehement and charged when we move out into society and hear the reaction of people to the killing of seals and dolphins, or to the vivisection of dogs. The first point I want to make today is that just because we are professionals in a cognate discipline, we do not own this issue in the same sense that we can expect to give the last word on questions well within our fields of special expertise. A large segment of the public knows what it thinks about the mental life of familiar animals, has a large stake in the practical consequences - hence laws and societies to prevent abuse, and stands ready to arbitrate the relevance of our conclusions. I do not mean that truth can be arrived at by popular vote. But we are asking about constructs (mind, mental life, feelings) that belong as much to the users of the English language as to scientific definition-makers. So, my point is that we should know what the public thinks, and I do not believe we do. It would be worthwhile even if the only consequence were that, like physicians and practical ecologists, we would prepare ourselves to speak to the public in a responsible way long before we feel we really know enough to be sure. But being so prepared would be a large step forward, in my view.

We know too little about the public's views, as distinct from the views of those who are vocal. Possibly there is a wide spectrum and a certain amount of inconsistency or uncertainty. We do our research and hold our discussions - both paid for by the public - in a social milieu unlike that in which we investigate disease or scientific questions with a noncontroversial aim. That means we have our impact on society in a charged atmosphere - and society has its impact on us. Especially on the issue of sentience and awareness, we are likely to be influenced in our judgement by what Bacon called the idols of the tribe and of the market place, since after all we are human, and we are reared in social groups.

I have tried a little public opinion polling, on samples of a few scores of people, in respect to the mental life of animals and in connection with the related question "animal rights." I am obviously an amateur at this tricky game and this is not the place to report in detail. However, I think it is germane to the struggles of this conference over which taxa have a mind, that people have a clear answer. On the average, with wide variance and some refusals to respond, people answer that mental life comes in degrees, as does intellectual capacity, awareness and capacity for suffering, and that animals differ widely in the degree or level of these attributes. In my small samples people rate chimpanzee above whale, dog above horse, pigeon above lizard, octopus above frog in respect to these attributes. The attributes are highly correlated. Subsets of respondents agree well, but I have not sampled a wide variety of people even within the U.S. society. I am not putting forward data or defending claims. It does seem appropriate in a conference like this to remind ourselves that much of the public has a position. I would also assert that the opinions of knowledgeable people are far from trivial in the domain of our stated purpose here, for the solid reason that we are not in a position to offer agreed-upon scientifically based ratings for the degree of awareness or the level of mental experience of nonhuman animals.

I return now to the converging contribution to our theme question of the views of various kinds of experts, such as are assembled here. We have a lot of relevant information and we have talked about it in our subgroup and plenary meetings. But we have been predictably timid about adding it up. We do not ask each other: "Taking all the relevant information you know, how would you scale the mental life of this list of animals?"

General readers of our book may well expect to find some consensus from this group about which animals have more and which have less mental experience - having read our stated goal - and they may feel disappointed by not finding such specifics. Perhaps this was what Bob Solomon was asking for;

if so, I agree. Jerry Fodor advises us to leave out the marginal cases until we have a theory for the paradigmatic cases. Unfortunately we cannot. We have waited for years for a theory of the human mind. Society demands and our own curiosity demands that we deal with animal minds without waiting. The very essence is: What about the marginal cases? Which species are marginal?

Our conference did struggle with this but some of the discussions essentially gave up and I think the reason is that we asked: "Which animals have mental experience and which do not?" instead of asking: "Which animals have more, and very roughly how much more?" I think we would have discovered some convergence of views among us - with a high divergence too, of course.

The consensus that I expect could be found among knowledgeable people will be blurred by two major factors, among others. One is that some think the cautious course is to assume that an animal has more awareness until evidence compels us to believe it has less. Others think the cautious position is to attribute awareness to animals only to the degree that evidence justifies such attribution. The difference is like that between "innocent until proven guilty" and "guilty until proven innocent." I bring this difference in standpoint out into the open in order to call attention to the asymmetry of these positions. Notice that the first position, assuming an attribute until it is disproven, is not based on any generally used scientific principle, let alone a necessary one, although some people who take this position clearly think it is the proper one, in terms of some kind of compelling parsimony. It is closer to an ethical norm in our culture. The second position, attributing a property only when evidence justifies it, is the one we use in physiology.

The other blurring factor is that some people believe there may be differences in kind or character of mental experience, making it more difficult to compare species with respect to degree of awareness of their respective kinds of experience.

In spite of these and other difficulties there are grounds for making preliminary estimates of the degree of awareness and its richness and complexity, and the repertoire, peaks, and valleys of affective experience in nonhuman animals, especially among the vertebrates. My main point should be clear by now, namely, that we should couch our questions and answers in terms of degree, and richness and repertoire and kinds, not in terms of presence or absence of a faculty difficult to define and even more difficult to detect at its earlier stages of evolution. At the same time let me underline a basic point illustrated by this conference, that although our theme question has a large subjective element, it can be studied scientifically.

The main help comes from the clinic where neurologists find that people are not simply conscious or unconscious but can be at any of several levels of awareness. Most of us have little knowledge of such states, but you can begin to empathize with them by comparing your experience of the unattended voices at the cocktail party, of the very hot serving plate when it is your own valuable heirloom plate, of the stages of falling asleep and waking up. The neurologist has a substantial, accumulated body of material that can tell us useful brain correlates of levels of awareness.

I spoke of the simplest kind of raw sensory perception. However, another set of structures is important for the addition of affect, associations of pleasure or pain, and evaluation for emotional overlay. We do not expect a high level of affective experience in a species that has little or no differentiated amygdala, septum, or prefrontal cortex, although the animal may react vigorously and adaptively as a spinal frog can do. We do not expect a high level of awareness in a species a) with very modest cortico-cortical connections, b) with too few neurons, especially excess neurons over a basic complement of sensory, motor, and visceral centers, c) which has little or no laminated cerebral cortex, or d) whose thresholds for psychoactive drugs are markedly higher than ours. All this means that while awareness and mental experiences of some kind are not denied altogether to nonhuman species, we can argue from many facts, inadequate as they are, that there is a marked gradation of these attributes within mammals and precious little of them in frogs and fish. The grounds for estimating mental levels of invertebrates are fewer but, apart from cephalopods, add up to substantially lower levels than fish even in the relatively advanced arthropods. The gastropods and other taxa are either similar or still lower than advanced arthropods. I would place the best developed cephalopods approximately equal to fish on the basis of the degree of differentiation of the brain and the limited ethology known.

The laws of some countries and the editorial policies of some journals treat all vertebrates alike, with respect to procedures required to avoid pain. This position may be compelled by social pressure but is incongruent both with my sample of public opinion and with a scientific appraisal of the graded basis for affective experience. The authors of these policies overlooked or underplayed a lot of evidence in adopting the position that every lamprey, shark, fish, frog, and snake has conscious feelings comparable to ours in degree and kind. Even to assume that all orders of mammals share our capacity for suffering in anything like a comparable degree is contrary to the relevant evidence - although, of course, this is far from denying that they have some level of feelings and awareness, anticipation and effect. How much more reasonable it would be to conclude from the available body of evidence, meager but not inconsiderable, that there is a profound gradation, as well as a specialization for different styles of life among the taxa, be they invertebrate, vertebrate, mammalian, or primate?

Although our conference report may disappoint the reader who looks for support for his favorite position or who expects a consensus of 50 experts as to which animals have a mind, we have achieved a modest goal. That is to point to a good many things that can be done to further understanding, by listing approaches and suggesting new research. This is where the workshop is pregnant and I congratulate you on the accomplishment. May your ideas gestate, come to term, and grow into lusty, self-conscious brain-children who will be fecund in their turn.

Gentle reader, judge not our accomplishment, in a fast, jet-lagged week in Berlin, by how much you find here of confident answers. Judge rather by how much it stimulates you to say "Ah, ha! That makes me think of a good experiment."

List of Participants

BEER, C.G.
Institute of Animal Behavior
Rutgers University
Newark, NJ 07102, USA

Field of research: Functional, developmental, and comparative study of communication behavior of gulls

BISCHOF, N.
Psychologisches Institut
Biol.-Mathematische Abteilung
8044 Zurich, Switzerland

Field of research: Human social motivation (proximate and ultimate causal analysis)

BLOOM, F.E.
The Salk Institute
P.O. Box 85800
San Diego, CA 92138, USA

Field of research: Cellular neuropharmacology of monoamine and peptide synapses

BROWN, R.H.
Dept. of Sociology
University of Maryland
College Park, MD 20742, USA

Field of research: Symbolism, language, and consciousness in relation to scientific and political rhetoric and practice

BULLOCK, T.H.
Dept. of Neurosciences
University of California
San Diego, A-001
La Jolla, CA 92093, USA

Field of research: Comparative neurophysiology

CERELLA, J.
V.A. Outpatient Clinic
Boston, MA 02108, USA

Field of research: Image processing in the pigeon

COHEN, R.J.
Fachgruppe Psychologie
Universität Konstanz, Postfach 5560
7750 Konstanz, F.R. Germany

Field of research: Cognitive impairments in aphasia, information processing in schizophrenia

CROOK, J.H.
Psychology Dept.
University of Bristol
Bristol 8, England

Field of research: Origin of consciousness theory, psychotherapy based upon western zen models of mind, social anthropology of village monastery relations in Himalayas, ethology

DAWKINS, M.
Animal Behaviour Research Group
Dept. of Zoology
University of Oxford
Oxford OX1 3PS, England

Field of research: Use of choice tests in assessment of animal welfare, perceptual aspects of searching in birds

DENNETT, D.C.
Dept. of Philosophy
Tufts University
Medford, MA 02155, USA

Field of research: Philosophy of mind, philosophy of psychology and artificial intelligence

DÖRNER, D.
Lehrstuhl Psychologie II
Universität Bamberg
8600 Bamberg, F.R. Germany

Field of research: Psychology of human thinking, problem-solving, planning and decision-making

DRENT, R.H.
Zoölogisch Laboratorium
Rijksuniversiteit Groningen
9751 Haren (Gr.), Netherlands

Field of research: Eso-ecology (birds)

ELEPFANDT, A.
Fakultät für Biologie
Universität Konstanz, Postfach 5560
7750 Konstanz, F.R. Germany

Field of research: Processing of spatial information in the lateral line system

FODOR, J.A.
Dept. of Psychology, E-10
M.I.T.
Cambridge, MA 02139, USA

Field of research: Philosophy of mind, psycholinguistics

GILLAN, D.J.
T23-1
General Foods Technical Center
Tarrytown, NY 10591, USA

Field of research: Experimental psychology

GLEITMAN, H.
Dept. of Psychology
University of Pennsylvania
Philadelphia, PA 19104, USA

Field of research: Memory in animals and humans, human cognitive processes, language

GOLDMAN-RAKIC, P.S.
Section of Neuroanatomy
Yale School of Medicine
New Haven, CT 06510, USA

Field of research: Developmental neurobiology, psychobiology

GOULD, J.L.
Dept. of Biology
Princeton University
Princeton, NJ 08544, USA

Field of research: Ethology

GRIFFIN, D.R.
The Rockefeller University
New York, NY 10021, USA

Field of research: Animal behavior, especially orientation behavior and cognitive ethology

HESSE, F.W.
Institut für Psychologie
der RWTH Aachen
5100 Aachen, F.R. Germany

Field of research: Cognitive psychology

HILLYARD, S.A.
Dept. of Neurosciences, M-008
University of California, San Diego
La Jolla, CA 92093, USA

Field of research: Neuropsychology, electrophysiology

HODOS, W.
Dept. of Psychology
University of Maryland
College Park, MD 20742, USA

Field of research: Comparative neuroanatomy, comparative psychology, animal psychophysics

KINTSCH, W.
Dept. of Psychology
Campus Box 346
University of Colorado
Boulder, CO 80309, USA

Field of research: Cognitive psychology

KLIX, F.
Sektion Psychologie
Humboldt-Universität
102 Berlin, German Democratic Republic

Field of research:

KLUWE, R.H.
Hochschule der Bundeswehr Hamburg
Fachbereich Pädagogik
Postfach 70 08 22
2000 Hamburg 70, F.R. Germany

Field of research: Development of executive control

KREBS, J.R.
Edward Grey Institute of
Field Ornithology
Dept. of Zoology
Oxford OX1 3PS, England

Field of research: Behavioral ecology

KUMMER, H.
Zoologisches Institut
Ethologie und Wildforschung
8050 Zürich, Switzerland

Field of research: Social relationships in nonhuman primates

LAMPRECHT, J.
Max-Planck-Institut für
Verhaltensphysiologie
8131 Seewiesen, F.R. Germany

Field of research: Attachment behavior in animals, socio-ecology of carnivores

LEVY, J.
Dept. of Behavioral Sciences
University of Chicago
Chicago, IL 60637, USA

Field of research: Asymmetry of the human brain

LINDAUER, M.
Zoologisches Institut (II)
der Universität Würzburg
8700 Würzburg, F.R. Germany

Field of research: Sociobiology, orientation in animals

LÜER, G.
Institut für Psychologie
der RWTH Aachen
5100 Aachen, F.R. Germany

Field of research: Cognitive psychology

LYNCH, G.S.
Dept. of Psychobiology
University of California
Irvine, CA 92717, USA

Field of research: Neuronal "plasticity"

MARKL, H.S.
Fakultät für Biologie
Universität Konstanz, Postfach 5560
7750 Konstanz, F.R. Germany

Field of research: Animal behavior

MARLER, P.R.
The Rockefeller University
Field Research Center
Millbrook, NY 12545, USA

Field of research: Animal communication, behavioral development

MASON, W.A.
California Primate Research Center
and Psychology Dept.
University of California
Davis, CA 95616, USA

Field of research: Comparative psychology

NEVILLE, H.J.
The Salk Institute
P.O. Box 85800
San Diego, CA 92138, USA

Field of research: Human neuropsychology

ORIANS, G.H.
Institute for Environmental Studies
and Dept. of Zoology
University of Washington
Seattle, WA 98195, USA

Field of research: Ecological aspects of animal social systems, evolutionary bases of human responses to plant shapes and landscapes

PARKER, G.A.
Dept. of Zoology
University of Liverpool, P.O. Box 147
Liverpool L69 3BX, England

Field of research: Evolutionary aspects of animal conflict, mating strategies and sexual selection

REITHER, F.
Lehrstuhl Psychologie II
Universität Bamberg
8600 Bamberg, F.R. Germany

Field of research: Human thinking and acting in complex situations (problem solving, cognitive psychology), metacognition

RISTAU, C.A.
The Rockefeller University
New York, NY 10021, USA

Field of research: Animal behavior/communication

ROZIN, P.
Dept. of Psychology
University of Pennsylvania
Philadelphia, PA 19104, USA

Field of research: Development of affect in humans, acquisition of food habits, evolution and nature of intelligence

SAVAGE-RUMBAUGH, E.S.
Yerkes Regional Primate Research Center
Emory University
Atlanta, GA 30322, USA

Field of research: Primate behavior, primate cognition and communication, early symbolic and linguistic processes

SEYFARTH, R.M.
The Rockefeller University
Field Research Center
Millbrook, NY 12545, USA

Field of research: Primate social behavior

SMITH CHURCHLAND, P.
Dept. of Philosophy
University of Manitoba
Winnipeg, Manitoba R3T 2N2, Canada

Field of research: Philosophy: theories of information processing

SOLOMON, R.C.
Dept. of Philosophy
University of Texas
Austin, TX 78712, USA

Field of research: Emotions, animal emotions

STADDON, J.E.R.
Dept. of Psychology
Duke University
Durham, NC 27706, USA

Field of research: Adaptive behavior and learning, theoretical analysis of operant behavior, behavioral ecology, experimental psychology

TERRACE, H.S.
418 Schermerhorn Hall
Columbia University
New York, NY 10027, USA

Field of research: Animal cognition: linguistic competence of apes, serial learning in pigeons

WEHNER, R.
Zoologisches Institut der Universität Zürich
8057 Zürich, Switzerland

Field of research: Vision in insects: neurophysiology and visually guided behavior

WELKER, C.
Zoologie und vergl. Anatomie
Gesamthochschule Kassel
3500 Kassel, F.R. Germany

Field of research: Social behavior of primates

Subject Index

Author Index

MIX
Papier aus verantwortungsvollen Quellen
Paper from responsible sources
FSC® C105338

If you have any concerns about our products,
you can contact us on
ProductSafety@springernature.com

In case Publisher is established outside the EU,
the EU authorized representative is:
Springer Nature Customer Service Center GmbH
Europaplatz 3, 69115 Heidelberg, Germany

Printed by Libri Plureos GmbH
in Hamburg, Germany